Why is the Universe Real?

From Quaternion & Octonion to Real Coordinates

Stephen Blaha Ph. D.
Blaha Research

Derivation of Real Space-Time and the Standard Model from QUeST
Derivation of QUeST from a One Dimension Basis BQUeST
Derivation of UTMOST from a One Dimension Basis BMOST
with a 10 Dimension Associated Space
Subspaces & Subtheories of QUeST
Subspaces & Subtheories of UTMOST
The Partition of Spaces
The Nature of Dimension with a Comment on Fractal Dimensions < 1
Factorization of UTMOST into TWO MOSTs
Factorization of MOST into TWO QUeSTs
Factorization of QUeST into REAL and IMAGINARY Space-times
Updated *Beneath the Quaternion Universe*

Pingree-Hill Publishing
MMXX

Rev. 00/00/01   August 14, 2020

To Margaret

Some Other Books by Stephen Blaha

All the Megaverse! Starships Exploring the Endless Universes of the Cosmos using the Baryonic Force (Blaha Research, Auburn, NH, 2014)

SuperCivilizations: Civilizations as Superorganisms (McMann-Fisher Publishing, Auburn, NH, 2010)

All the Universe! Faster Than Light Tachyon Quark Starships & Particle Accelerators with the LHC as a Prototype Starship Drive Scientific Edition (Pingree-Hill Publishing, Auburn, NH, 2011).

Unification of God Theory and Unified SuperStandard Model THIRD EDITION (Pingree Hill Publishing, Auburn, NH, 2018).

The Exact QED Calculation of the Fine Structure Constant Implies ALL 4D Universes have the Same Physics/Life Prospects (Pingree Hill Publishing, Auburn, NH, 2019).

Unified SuperStandard Theory and the SuperUniverse Model: The Foundation of Science (Pingree Hill Publishing, Auburn, NH, 2018).

Quaternion Unified SuperStandard Theory (The QUeST) and Megaverse Octonion SuperStandard Theory (MOST) (Pingree Hill Publishing, Auburn, NH, 2020).

Unified SuperStandard Theories for Quaternion Universes & The Octonion Megaverse (Pingree Hill Publishing, Auburn, NH, 2020).

The Essence of Eternity: Quaternion & Octonion SuperStandard Theories (Pingree Hill Publishing, Auburn, NH, 2020).

Available on Amazon.com, bn.com Amazon.co.uk and other international web sites as well as at better bookstores (through Ingram Distributors).

CONTENTS

FIGURES and TABLES

Beneath the Quaternion Universe

INTRODUCTION

The Quaternion Unified SuperStandatd Theory (QUeST) is a theory for our universe that is based on a 32 complex quaternion space. It implies the Unified SuperStandard Theory (UST) of the author to the author's initial surprise. The Megaverse theory UTMOST (Megaverse Octonion SuperStandard Theory) is based on a 64 complex octonion space. Both theories can be based on deeper one dimension theories. We show that in later in the updated *Beneath the Quaternion Universe* book included here for the reader's convenience. References to material in that book are denoted "Beneath."

The primary new interest in this book is the description of subspaces and subtheories of UTMOST and QUeST. QUeST, MOST, and UTMOST spaces can be partitioned into subspaces with their own theories. Perhaps the most immediately interesting subspaces are a REAL subspace and IMAGINARY subspace of QUeST. We find their space-times have 3+1 dimension real, and 3+1 dimension imaginary coordinates respectively. The REAL subspace gives The Standard Model of Elementary Particles. QUeST contains this subspace and theory. QUeST has a richer set of dimensions, symmetries, and particles.

UTMOST, which appears to be the correct theory of the Megaverse, has a complex structure and set of subspaces. The subspaces include MOST and QUeST subspaces.

The extraction of subspaces from QUeST and UTMOST follow certain rules that are provided in chapter 2. The key feature of these rules is a requirement on theories such as QUeST and UTMOST to be ultimately based on a one dimension – one fermion substratum that we have called BQUeST and BMOST respectively.

After considering QUcST and UTMOST we consider the nature of dimension with a comment on fractal dimensions < 1, which do not seem to have been previously considered. We show that all definitions of dimension appear to be operational and not based on more primitive terms. Thus we are led to believe that QUeST and UTMOST are at the deepest level of Physical Reality.

The reader may wish to read the Introduction to Beneath provided later. It extends the discussion of QUeST and UTMOST features.

1. Summary of UST, QUeST, MOST, UTMOST

1.1 Some Rules for Determining the Form of the Theories

The previous books have developed a number of "rules" that shape the form of the dimension and fermion arrays.

A. In the study of the derivation of the theories from their one dimension – one fermion basis in a two-step iteration from 1×1 arrays to 4×4 arrays and then to the complete dimension array, we found that the first step suggested *4 ×4 blocks (reflecting a U(8) structure)* within each of the theories' dimension arrays and fermion spectrum. The 4×4 blocks led to having U(2) Dark groups and U(4) Layer groups in the theories.

B. The dimension arrays should be made into square arrays if possible. This reordering does not change the arrays' physical implications.

1.1 UST and QUeST

UST and QUeST have the same set of internal symmetries and fundamental fermion spectrum.[1]

The major features of QUeST Space follow.

QUeST Space: Thirty-two complex quaternion dimensions totaling 256 dimensions
Layers: Four layers with each consisting of 8 complex quaternions
QUeST and UST Internal Symmetry group:[2]

$$[SU(2) \otimes U(1) \otimes SU(3) \otimes SU(2) \otimes U(1) \otimes SU(3) \otimes U(4)^4 \otimes U(2)]^4 \qquad (1.1)$$

---

[1] See Blaha (2020c) for UST details. The surprising similarity of QUeST to UST is encouraging.
[2] The U(2) groups are Dark groups mixing Normal and Dark matter.

QUeST Space-Time: 3+1 complex quaternion dimensions
QUeST Dimensions Array:

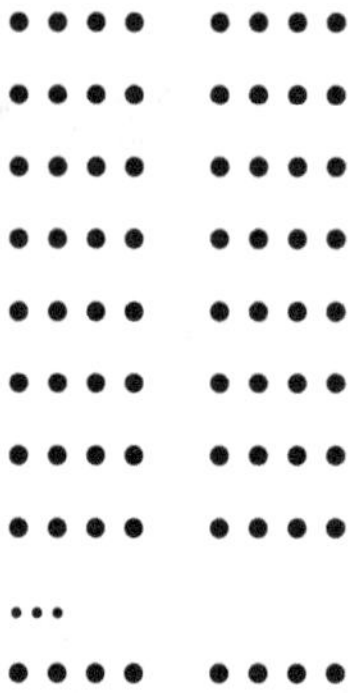

```
• • • •   • • • •
• • • •   • • • •
• • • •   • • • •
• • • •   • • • •
• • • •   • • • •
• • • •   • • • •
• • • •   • • • •
• • • •   • • • •
• • •
• • • •   • • • •
```

Figure 1.1. The 32 complex quaternion dimensions QUeST array.

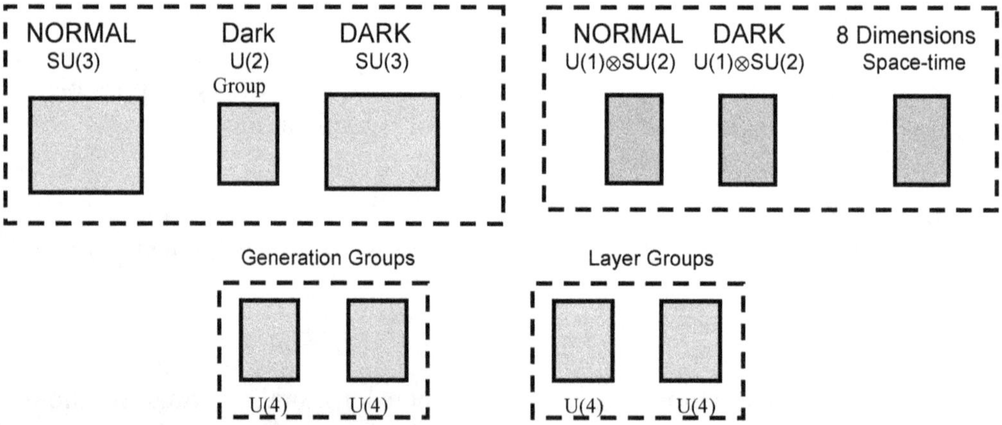

Figure 1.2. Block array of the internal symmetry groups for *one* QUeST layer, which consists of 8 complex quaternion dimensions. Each "dashed" block (regardless of apparent size) contains 4×4 = 16 dimensions. The set of 4 blocks contain $4 \times 16 = 64$ dimensions. The Dark U(2) group supports transformations (rotations) between Normal and Dark matter.

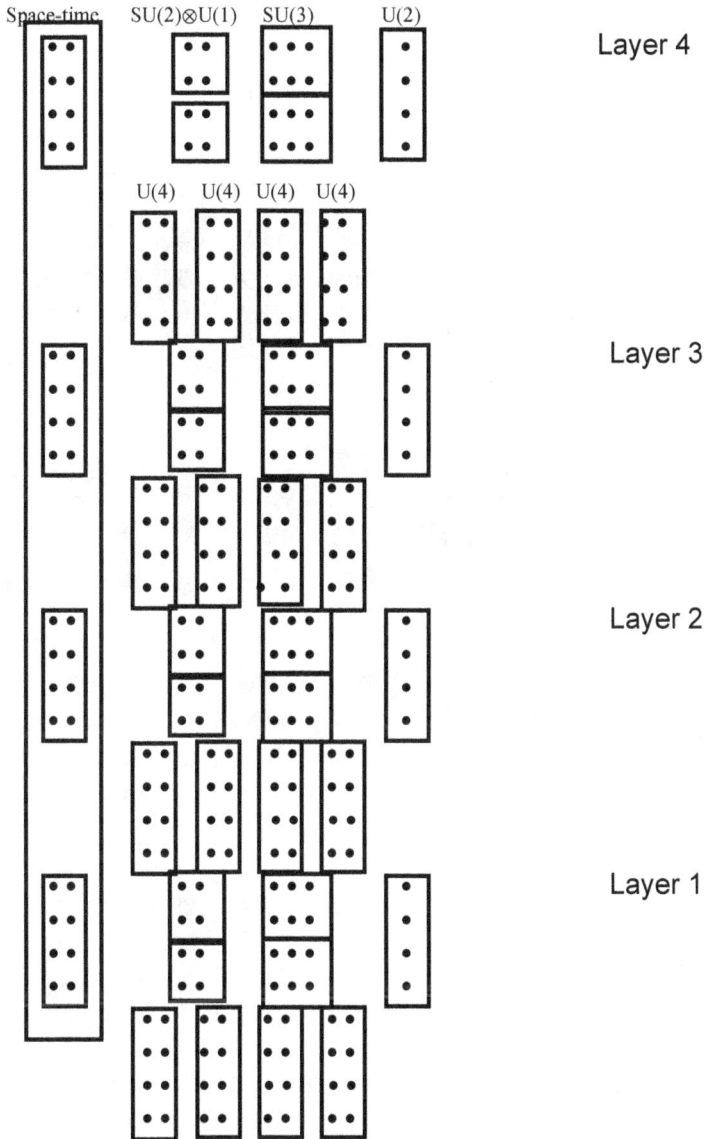

Figure 1.3. The four layers of QUeST internal symmetry groups (and space-time) for 32 dimension complex quaternion space. Note: each row has an 8 •

complex quaternion. Note the left composite blocks combine to specify a 4 dimension complex quaternion space-time.

QUeST and UST Fermion Arrays:

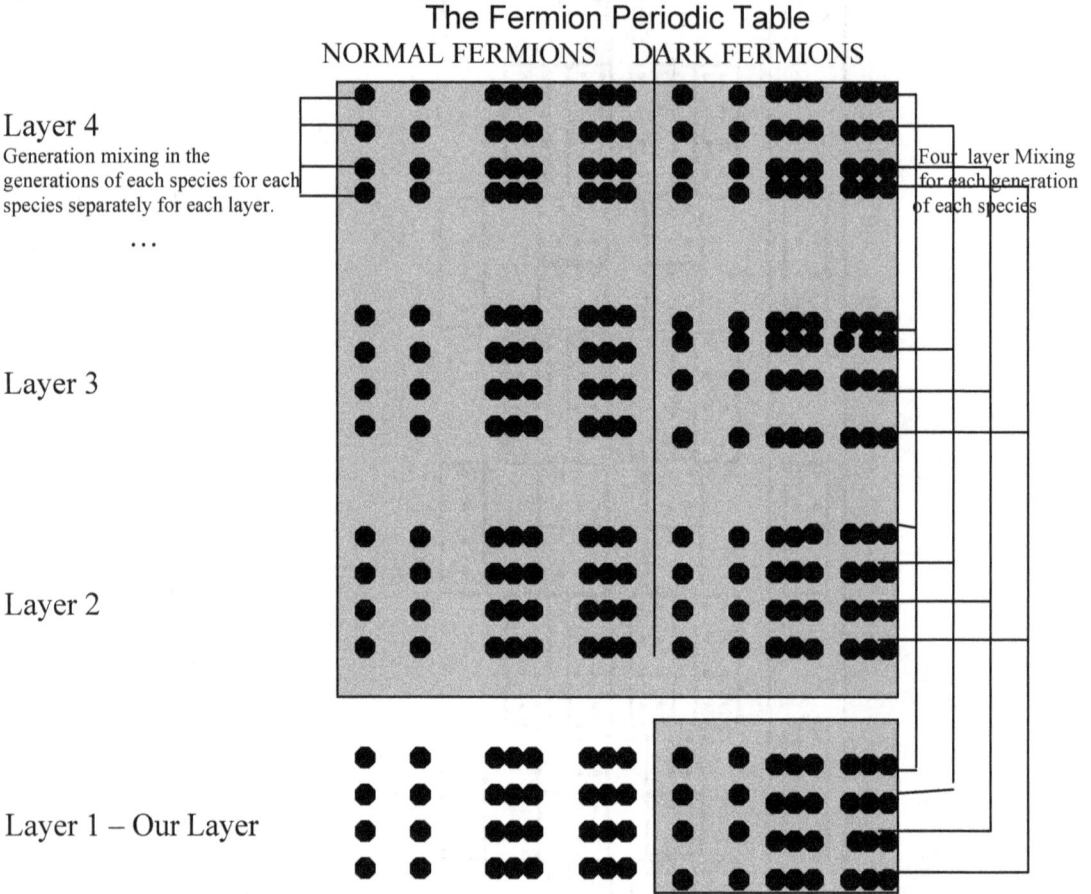

The Fermion Periodic Table

NORMAL FERMIONS   DARK FERMIONS

Layer 4
Generation mixing in the
generations of each species for each
species separately for each layer.

Four layer Mixing
for each generation
of each species

...

Layer 3

Layer 2

Layer 1 – Our Layer

Figure 1.4. Fermion particle spectrum and partial example of pattern of mass mixing of the Generation group and of the Layer group. Unshaded parts are the known fermions with an additional, as yet not found, 4th generation. The lines on the left side (only shown for one layer) display the Generation mixing within each layer's species. The Generation mixing applies within each layer using a separate Generation group for each layer. The lines on the right side show Layer group mixing with the mixing amongst all four layers for each of the four generations individually. There are four Layer groups. There are 256 fundamental fermions. QUeST and UST have the same fermion spectrum.

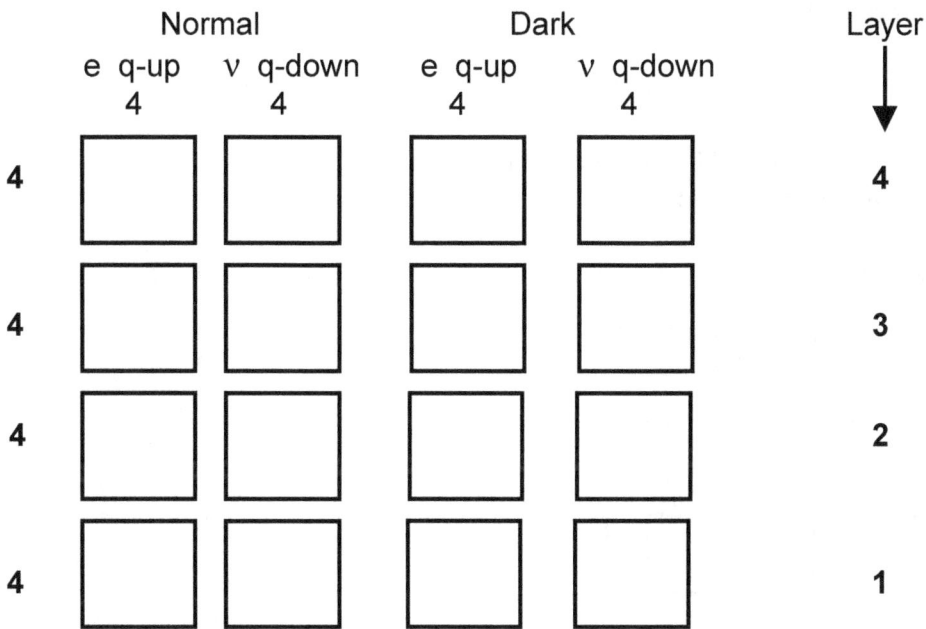

Figure 1.5. Block form of 16 × 16 QUeST fermion array with each block row corresponding to one layer. Each block contains four generations of four fermions (reordered). The result is 4 × 4 blocks. In the top label: e q-up indicates a charged lepton – up-type quark pair, v q-down indicates a neutral lepton – down-type quark pair, and so on. 256 fermions.

1.2 MOST

The major features of MOST Space follow.

MOST Space: Thirty-two complex octonion dimensions totaling 512 dimensions. A 32 × 16 array of dimensions.

Layers: Four layers with each consisting of 8 complex octonions (128 dimensions per layer)

MOST Internal Symmetry group:[3]

$$[[SU(2)\otimes U(1)\otimes SU(3)]^4 \otimes U(4)^8 \otimes U(2)^2]^4 \qquad (1.1)$$

MOST Space-Time: 7+1 complex quaternion dimensions
MOST Dimension Array:

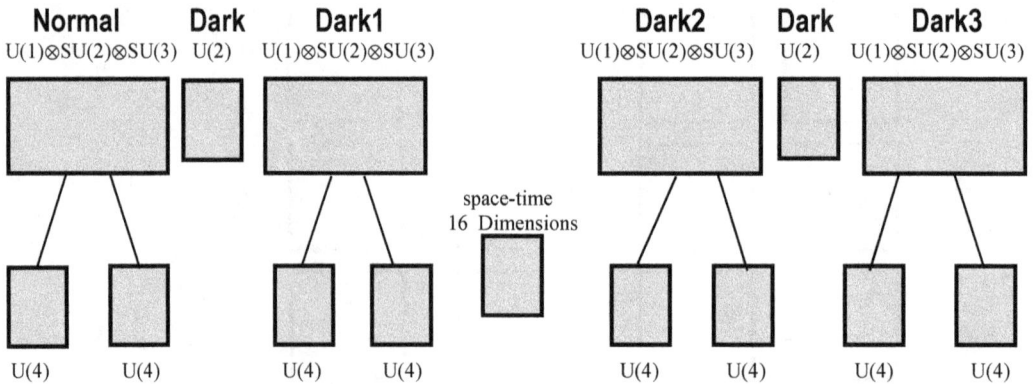

Figure 1.6. The internal symmetry groups of *one* MOST layer with two Dark U(2) groups. The lower U(4) groups are the Generation and Layer number groups. One pair of each number group is for each of the four U(1)⊗SU(2)⊗SU(3) factors above. One Dark U(2) group mixes Normal and

---

[3] The U(2) groups are Dark groups mixing Normal and Dark matter.

Dark1. The other U(2) group mixes Dark2 and Dark3.There are 128 dimensions.

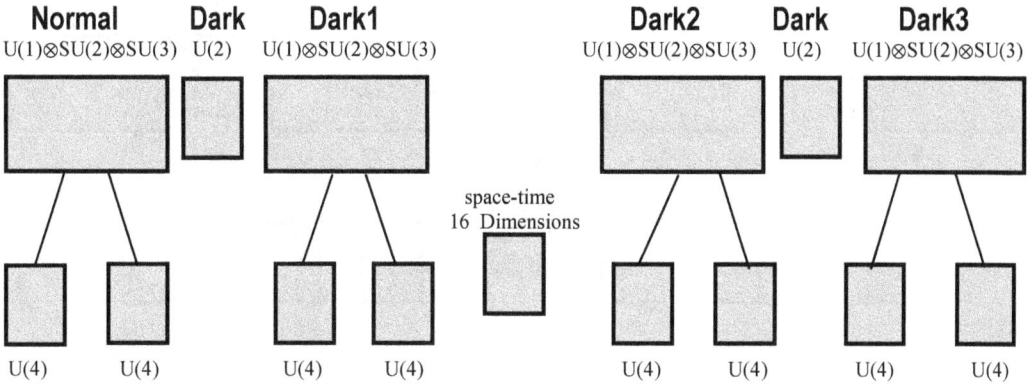

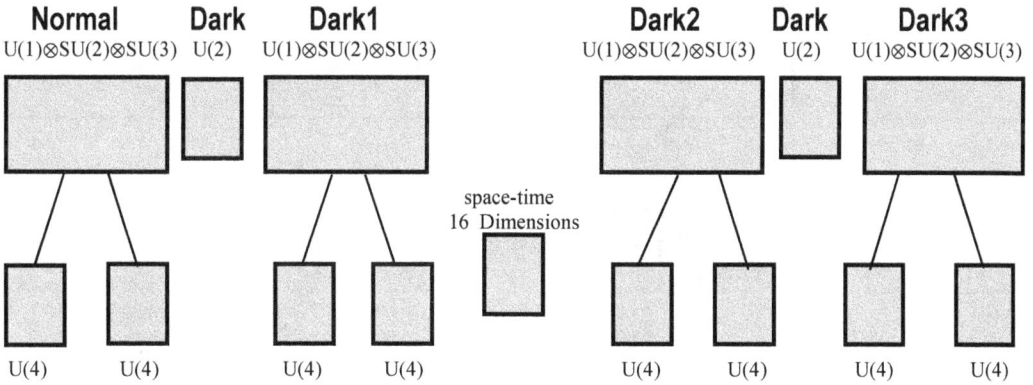

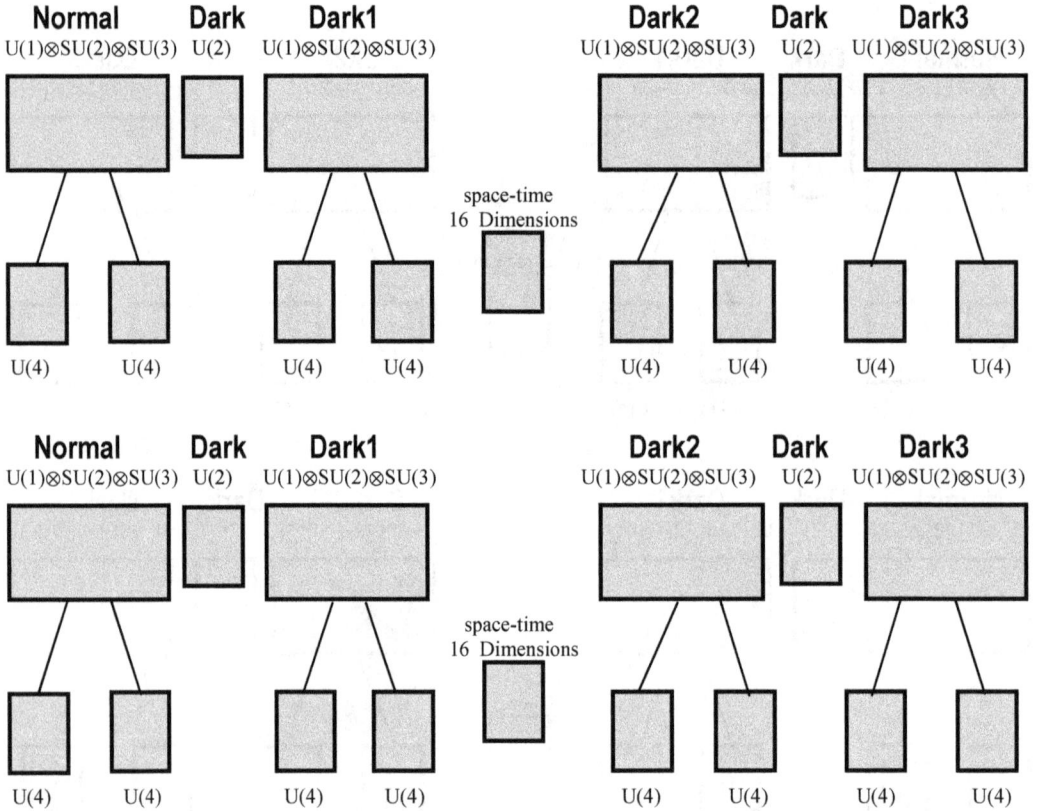

Figure 1.7a. The internal symmetry groups of *four* layer MOST with two Dark U(2) groups. One Dark U(2) group mixes Normal and Dark1. The other U(2) group mixes Dark2 and Dark3.There are 128 dimensions. The 16 dimension space-time parts of the four layers total to make a 7+1 complex quaternion space-time.

Normal + Dark1   **Dark2 + Dark3**

Figure 1.7b. 4 × 4 blocks within the four block, 8 × 8 sections for each pair of Normal and Dark sets of dimensions. They form the 16 × 32 = 512 MOST dimension array. See Fig. 1.2 for the makeup of each 8 × 8 section.

MOST Fermion Array:

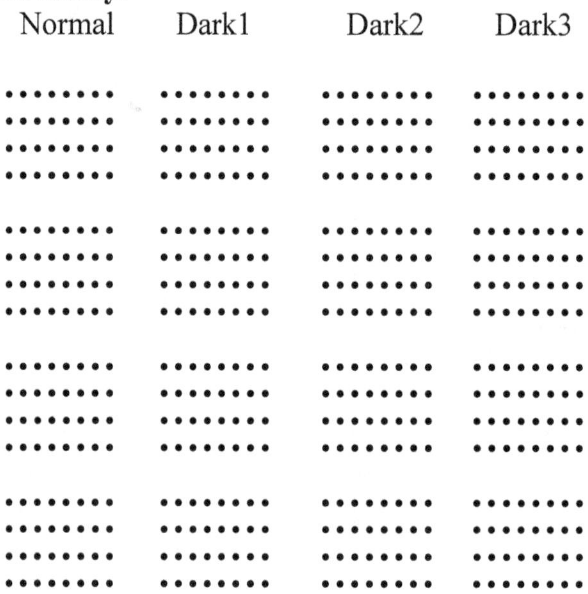

| | Normal | Dark1 | Dark2 | Dark3 |

Figure 1.8. Spectrum of the fermions of MOST. Each fermion is represented by a •. Quark triplets are represented by a single •. Four sets of four species in four generations which are in turn in 4 layers. There are 512 fundamental fermions taking account of quark triplets. Note the Layer groups determine the layers in QUeST and UST. **They require 4 layers of 8 complex octonions in Megaverse space leading to the 32 dimension complex octonion space.**

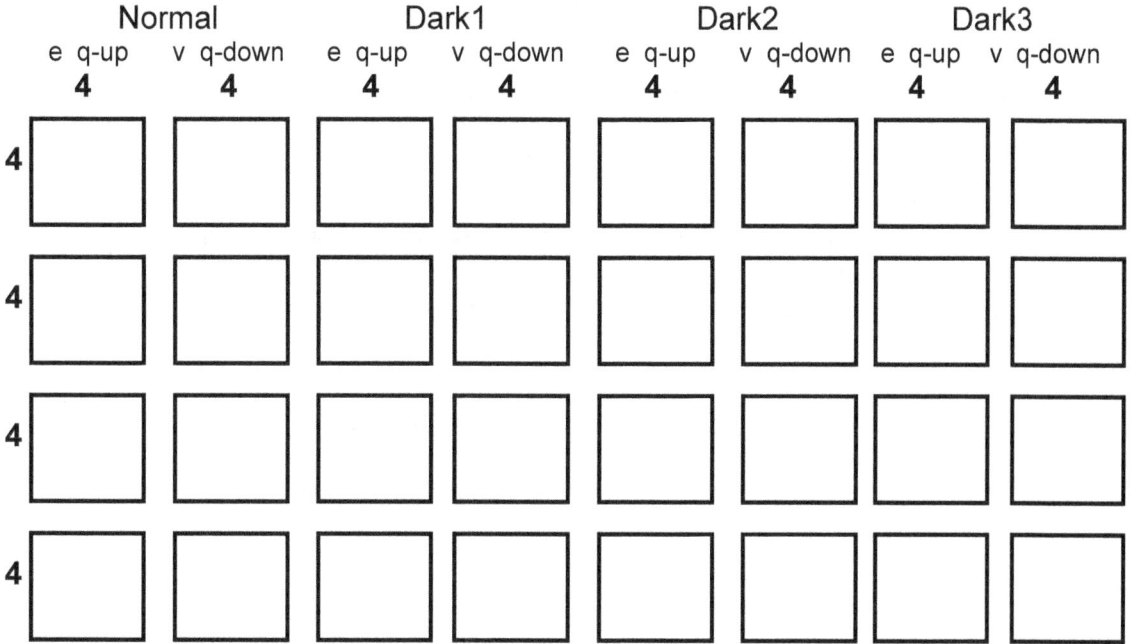

Figure 1.9. Block form of MOST fermion array with each block row corresponding to one layer. Each block contains four generations of fermions. The result is 4 × 4 blocks. The fermion rows are reordered. The label e q-up indicates a charged lepton – up-type quark pair, v q-down indicates a neutral lepton – down-type quark pair, and so on. *This form explains why generations come in fours.*

1.3 UTMOST

UTMOST Space: - Two Formats

The first format seems more "natural." The second format is better adapted to derive UTMOST from BMOST and to reveal the 16 × 16 block formatting. We will use both formats in the following figures.

1. Sixty-four complex octonion dimensions totaling 1024 dimensions A 64 × 16 array of dimensions.
   Layers: Eight layers with each consisting of 8 complex octonions (128 dimensions)

2. Thirty-two double complex octonion dimensions totaling 1024 dimensions A 32 × 32 array of dimensions.
   Layers: Four layers with each consisting of 8 double complex octonions (128 dimensions)

UTMOST Internal Symmetry group:[4]

$$[SU(2)\otimes U(1)\otimes SU(3)\otimes U(4)^4\otimes U(2)]^{16} \tag{1.1}$$

UTMOST Space-Time: 7+1 complex octonion dimensions

UTMOST Dimension Array

The UTMOST dimension array has the same form as that of MOST but with with complex octonion rows and twice the number of rows: from 32 to 64. Thus Fig. 1.6 describes the subgroups in the each of the 8 layers of UTMOST.

The UTMOST dimension array is 64 × 16 = 1024 dimensions. In deriving UTMOST from the BMOST one dimension array, we found it necessary to restructure the array to a 32 × 32 format. The reordered array better brings out the 16 × 16 block substructure. One can reorder by moving the lower 32 rows, up and to the right of the

---

[4] The U(2) groups are Dark groups mixing Normal and Dark matter.

top 32 rows. See Fig. 1.10 for the resulting UTMOST dimension pattern. The moved four rows are now Dark4, Dark5, Dark6 and Dark7.

Figure 1.10. The 64 complex octonion dimension UTMOST array.

Figure 1.11. The reordered UTMOST array with 32 dimensions and *double* complex octonion rows. It has a 32 × 32 form with 1024 dimensions.

	Normal + Dark1		Dark2 + Dark3		Dark4 + Dark5		Dark6 + Dark7	
	4	4	4	4	4	4	4	4
4								
4								
4								
4								
4								
4								
4								
4								

Figure 1.12. Four layers in 32 × 32 dimension array of 4 × 4 blocks, within the four block 8 × 8 sections for each pair: Normal+Dark1, Dark2+Dark3, Dark4+Dark5 and Dark6+Dark7. In total they form the 32 × 32 = 1024 UTMOST dimension array. See Fig. 1.2 for the makeup of each 8 × 8 section.

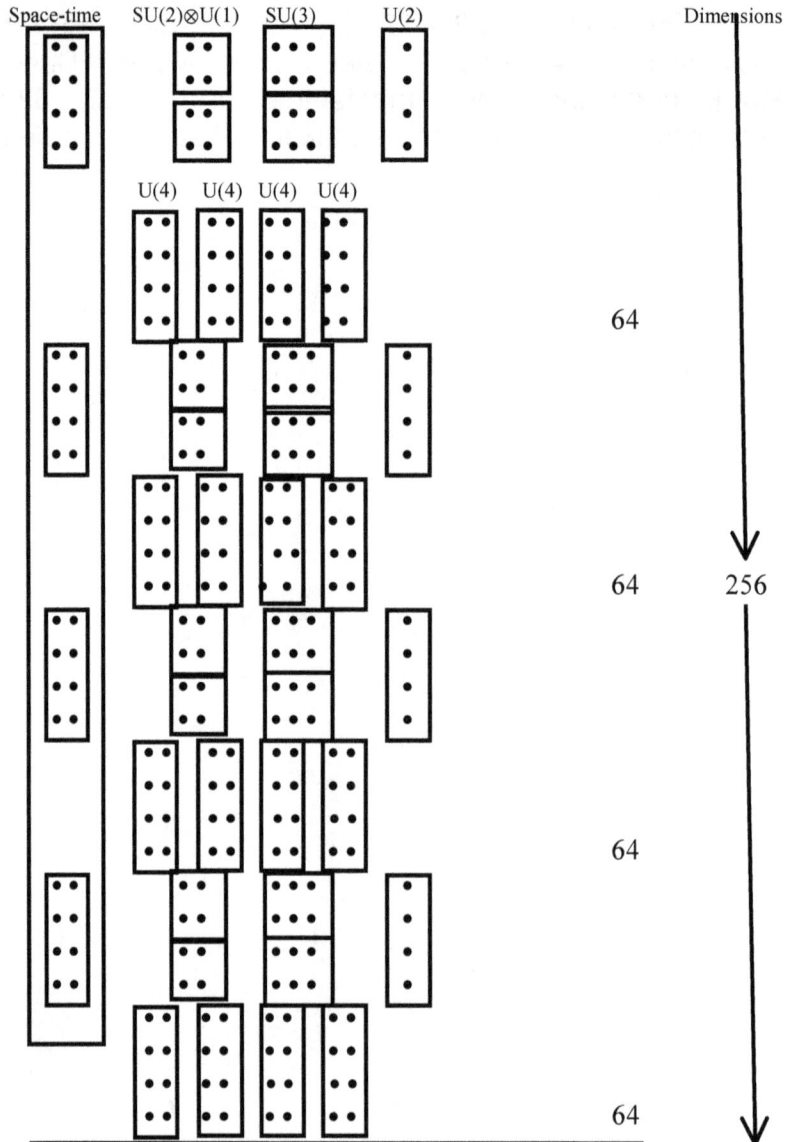

Figure 1.13 Two of the eight rows in Fig. 1.12 for one of the four layers of UTMOST dimensions.

UTMOST Fermion Array

Normal	Dark1	Dark2	Dark3	Dark4	Dark5	Dark6	Dark7

Figure 1.14. Spectrum of UTMOST fermions in a 16×64 format. Each fermion is represented by a •. Including each quark. Each set of eight •.'s represents a charged lepton, a neutral lepton, three up-type quarks, and three down-type quarks. There are eight sets of four species in four generations which are in turn in 4 layers. There are 1024 fundamental fermions taking account of quark triplets. Note: Quark singlets won't do; triplets are required.

	Normal				Dark1				Dark2				Dark3	
	e q-up	v q-down			e q-up	v q-down			e q-up	v q-down			e q-up	v q-down
	4	4			4	4			4	4			4	4
4														
4														
4														
4														

	Dark4			Dark5			Dark6			Dark7	
4											
4											
4											
4											

Figure 1.15. Block form of the 32 × 32 UTMOST fermion array with each row corresponding to *half of a layer*. (Compare to Fig. 1.12.) Thus 8 × ½ = 4 layers results. Each block contains four generations of fermions. The result is sixty-four 4 × 4 blocks. The label e q-up indicates a charged lepton – up-type quark pair, v q-down indicates a neutral lepton – down-type quark pair, and so on. *The form displayed here explains why generations come in fours.*

1.4 Connections between QUeST, BQUeST, UTMOST, and BMOST

In Beneath (and earlier) we found that one could construct QUeST from a one-dimension, one fermion[5] basis, called BQUeST, with the fermion existing in an 8 dimension space-time. Now MOST has an 8 complex quaternion space-time; UTMOST has an 8 complex octonion space-time. Therefore we suggested that a BQUeST/QUeST universe begins in a MOST or an UTMOST Megaverse.

In Beneath we also found that one could construct UTMOST from a one-dimension, one fermion[6] basis, called BMOST, with the fermion existing in a 10 dimension space-time. Since BMOST/UTMOST does not apparently reside in a larger space. We concluded that there is a 10 dimension space-time within which BMOST/UTMOST is generated. Since 10 dimensions appear in Superstring theories, there is a possibility of a connection between BMOST/UTMOST and Superstring theories.

---

[5] The fermion had two fields in a PseudoQuantum formulation. See Blaha (2020c) for a description of Blaha's PseudoQuantum field theory formulation.
[6] The fermion had two fields in a PseudoQuantum formulation. See Blaha (2020c) for a description of Blaha's PseudoQuantum field theory formulation.

2. Partitioning of Spaces

In Beneath we discussed the partition of the UTMOST space into two MOST subspaces based on the requirement that the UTMOST dimension array consisted of 64 U(8) blocks of 16 dimensions . We found that we had to replace the U(4) Dark groups with U(2) Dark groups – one set for each of the layers of UTMOST. We separated the UTMOST dimensions into two subspaces. Each subspace was a MOST space.

In this chapter we state the rules for determining subspaces. In the following chapters we will consider UTMOST and MOST subspaces.

2.1 Some Rules for Determining the Form of the Theories

The previous books have developed a number of "rules" that shape the form of the dimension and fermion arrays. The rules are:

A. The array must be derivable from a fundamental one dimension- one fermion theory.

B. To establish a derivation of quaternion and octonion theories from one dimension with one fermion, the dimension array must be a square array or transformable to square array.[7]

C. Since the array must be generated from the independent spinor components of the off-shell fermion, we found the allowed array width (and length) is $D_A = 2^{D/2}$ where $D = 8$ and 10 based on the number of components of a spinor. Thus physically acceptable arrays are 16×16 with 256 dimensions, or 32×32 with 1024 dimensions.

---

[7] The reordering of a dimension array to make it square does not affect the physical consequences.

D. In the study of the derivation of theories[8] from their one dimension – one fermion basis in a two-step iteration from 1×1 arrays to 4×4 arrays and then to the complete dimension array, we found that the first step suggested *4 ×4 blocks (reflecting a relic U(8) substructure)* within each of the theories' dimension arrays and fermion spectrum. The separation into 4×4 blocks required U(4) Dark groups is separated into U(2) factors. U(4) → U(2)⊗U(2). It is also based on Layer groups being U(4) groups.

E. The 4×4 blocks appear in *superblocks*, which consist of four blocks. Superblocks contain 64 dimensions. Thus a superblock can be viewed as the fundamental representation of (broken) U(32). Each theory must contain a set of superblocks. A theory with a 16×16 dimension array has four superblocks. A theory with a 32×32 dimension array has sixteen superblocks.

2.2 Extended Rules

The rules in section 2.1 support a derivation of a theory from a 1 dimension space by attributing the one jump to the size of the theory's array to the fermion's set of spinor components. As pointed out in Beneath one can reach the size of an array in two jumps. For example the 32×32 UTMOST array can be generated from the fermion having four independent spinor components, each of which becoming a fermion with eight spinor components.[9] The result is a 32×32 array for UTMOST.

Thus items C and D above can be generalized to the two step jump. One immediate consequence of this extension for the case of a 4-jump followed by a 6-jump is a 24×24 array. This array could be viewed as describing a 24 trioctonion space.

We will not consider this possibility here since QUeST and UTMOST seem to well describe the universe and Megaverse respectively.

---

[8] See Beneath.
[9] We find it more comfortable to consider just the fermion functionals with their spinor components at each step.

3. QUeST Subspaces

3.1 Partition of QUeST into Two Subspaces

Section 1.1 contains figures describing the dimensions and fermion spectrum of QUeST. The dimension array of QUeST can be partitioned into two 128 dimension parts. Figure 1.2 is separated into two parts. It appears reasonable to identify the parts as Normal and Dark since normal matter is known to be distinct from Dark matter.

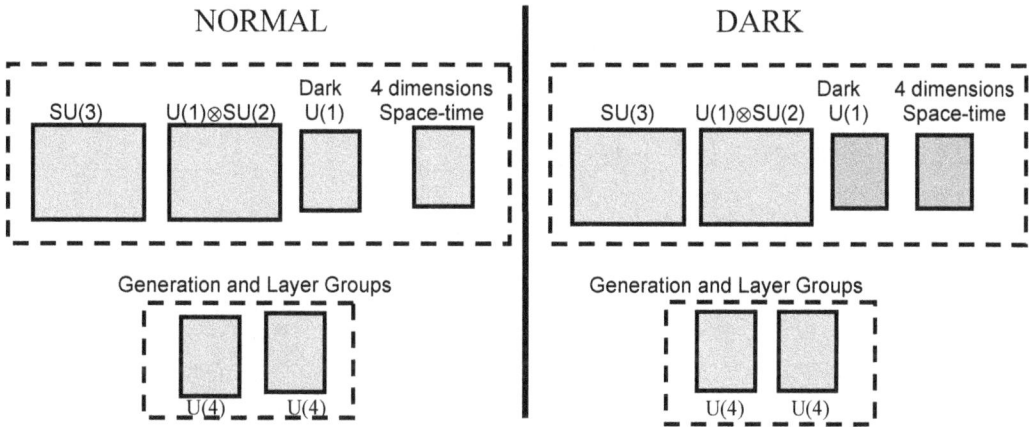

Figure 1.2s. Separated parts of one layer of QUeST. The other three layers are copies. The total number of the dimensions in each part is 4×32 = 128. The internal symmetry groups are identified. Each "dashed" block contains 4×4 = 16 dimensions. Each layer in each part contains 32 dimensions. The Dark U(1) group supports transformations (rotations) of fermions in each part. Note the Dark U(2) group in Fig. 1.2 is split into U(1)⊗U(1) to achieve separation.

Figure 1.4 and 1.5 show a clear separation of the fermion spectrum into two parts.

Thus the subspaces cleanly separate the Normal and Dark sectors of QUeST. They provide a possible justification for the non-interaction of Normal and Dark matter.

The separation of the QUeST dimension array into the above subspaces causes each part to be a 32 quaternion subspace. Each part has a four quaternion dimensions space-time. Apparently QUeST's complex quaternion coordinates combine the Normal and Dark sectors. See Fig. 1.1s below.

```
• • • •
• • • •
• • • •
• • • •
  • • •
• • • •
```

Figure 1.1s. A 32 quaternion dimension array.

4. A Further Partition into Real and Imaginary 3+1 Dimension Space-times

We can further partition each part into real and imaginary 3+1 dimension space-times.

REAL SPACE-TIME

NORMAL   DARK

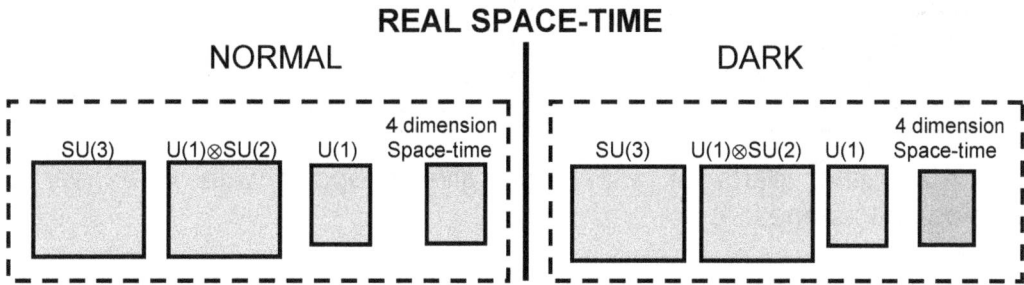

Figure 4.1. Separation of each part to obtain subparts with real 3+1 dimension space-times. The Dark, Generation and Layer groups are absent. Each 'dotted" block has 16 dimensions within. There is only one layer. Note the SU(3) and U(1)⊗SU(2) groups are the same in the REAL Normal and IMAGINARY Normal sectors. As a result the REAL Normal and IMAGINARY Normal fermions interact with each other. IMAGINARY Normal fermions are "part" of the REAL Normal fermion set. Similarly IMAGINARY Dark fermions are "part" of the REAL Dark fermion set.

In UST and QUeST the four species of fermions (charged lepton, neutral lepton, up-type quark, and down-type quark) were divided into Dirac-type particles (charged lepton, and up-type quark) and tachyon-type particles (neutral lepton, and down-type quark) (See Fig. 1.5.)

If we follow that procedure here and realize that Fig. 4.1, which is only for real-valued coordinates, is only half the story—the other halves are for imaginary-valued coordinates, then we must include Fig. 4.2 for imaginary coordinates.[10]

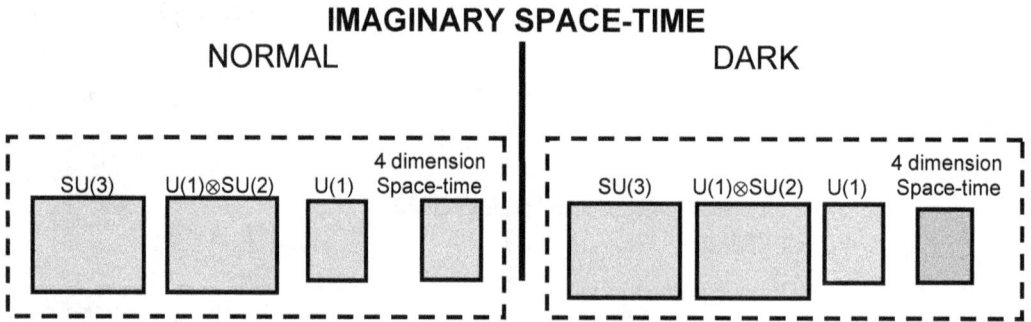

IMAGINARY SPACE-TIME

NORMAL | DARK

| SU(3) | U(1)⊗SU(2) | U(1) | 4 dimension Space-time | | SU(3) | U(1)⊗SU(2) | U(1) | 4 dimension Space-time |

Figure 4.2. Other parts of separation give subparts with imaginary 3+1 dimension space-times.

The fermions of Normal and Dark fermions appear in rows of four fermions for each Normal and Dark, and for Real and Imaginary parts. In the absence of a Generation group we choose four generations. If the number of dimensions of each part (16) equals the number of fermions, then there are two generations. The fermions for the REAL subspace appear in Fig. 4.3.

---

[10] The four species of coordinates are derived using four Complex Lorantz group boosts. See Blaha (2020c).

REAL

NORMAL	DARK
e q-up	e q-up
• • • •	• • • •
• • • •	• • • •
• • • •	• • • •
• • • •	• • • •

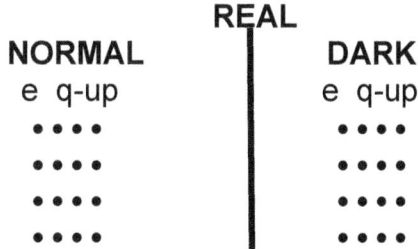

Figure 4.3. Normal and Dark fermions for Real-valued coordinates subspace. One charged lepton and three up-type quarks per generation for Normal and Dark.

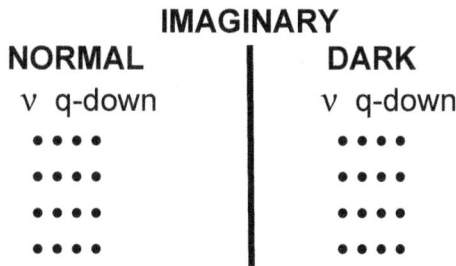

IMAGINARY

NORMAL	DARK
ν q-down	ν q-down
• • • •	• • • •
• • • •	• • • •
• • • •	• • • •
• • • •	• • • •

Figure 4.4. Normal and Dark fermions for Imaginary-valued coordinates subspace. One neutral lepton and three down-type quarks per generation for Normal and Dark.

Note that Figs. 4.3 and 4.4 have 16 fermions in Normal and Dark for Real and Imaginary coordinates. The number of fermions (16) in each subspace equals the number of dimensions (16).

4.1 Physics of the Partitioned QUeST into Real and Imaginary Parts

Above we have further partitioned QUeST into REAL and IMAGINARY parts. If we take the REAL subspace as reality, then we must ask what its components are. First we have the real 4-dimension coordinate system. Then we note the Normal SU(3) and U(1)⊗SU(2) internal symmetry groups are the same in both the REAL and

IMAGINARY parts. Similarly, the SU(3) and U(1)⊗SU(2) internal symmetry groups are the same in both the REAL and IMAGINARY Dark parts.

Since they support interactions among all fermions charged leptons, up-type quarks, neutral leptons, and down-type quarks, all fermions of the IMAGINARY Normal and Dark parts types are present in the REAL subspace in the respective Normal and Dark parts.

The IMAGINARY neutral leptons, and down-type quarks are in the REAL Normal sector due to the common internal symmetry groups of the REAL and IMAGINARY Normal sectors. The Dark REAL and IMAGINARY symmetry groups, although different from the Normal symmetry groups, are the same. Thus the Dark IMAGINARY fermions are part of the REAL Dark fermion set.

The depiction above corresponds exactly to the derivation of the four fermion species from Complex Lorentz boosts in UST in Blaha (2007b) through (2020c). Thus neutral leptons and down-type quarks are tachyonic due to the "imaginary" boosts of the Complex Lorentz group, and the charged leptons and up-type quarks are Diracian due to "real" Lorentz boosts.

From Fig. 4.1 we have achieved REAL 4 dimension space-time. However the UST and QUeST theories have a more complete characterization of the framework of elementary particle physics. Thus they are the theories of choice.

5. UTMOST Subspaces

UTMOST[11] can be partitioned into several levels of subspaces. We modify the UTMOST figures of chapter 1 to show the partition into two subspaces. These spaces will be MOST spaces. Then we partition a MOST subspace into two QUeST subspaces. The partitioned UTMOST dimension arrays are:

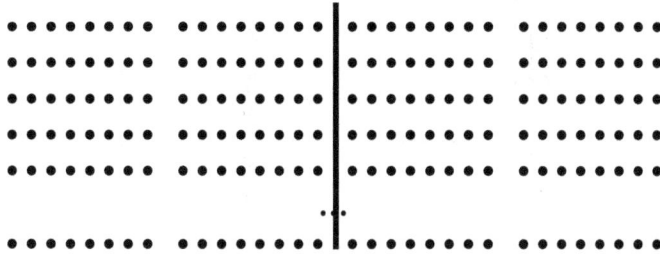

Figure 1.11s. The partition of the reordered 32 dimension UTMOST array The size of each subspace is 32×16 = 256 dimensions.

---

[11] We will consider MOST subspaces below.

Normal + Dark1		Dark2 + Dark3		Dark4 + Dark5		Dark6 + Dark7	
4	4	4	4	4	4	4	4
4							
4							
4							
4							
4							
4							
4							
4							

Figure 1.12s. Partition of four layers in 32 × 32 dimension array of 4 × 4 blocks, within the four block with 8 × 8 sections for each pair: Normal+Dark1, and Dark2+Dark3 form one subspace. Dark4+Dark5 and Dark6+Dark7 form the other subspace. In total each subspace is a 32 × 16 = 512 dimension array. See Fig. 1.2 for the makeup of each 8 × 8 section. The dimensions of each subspace correspond to the MOST array space.

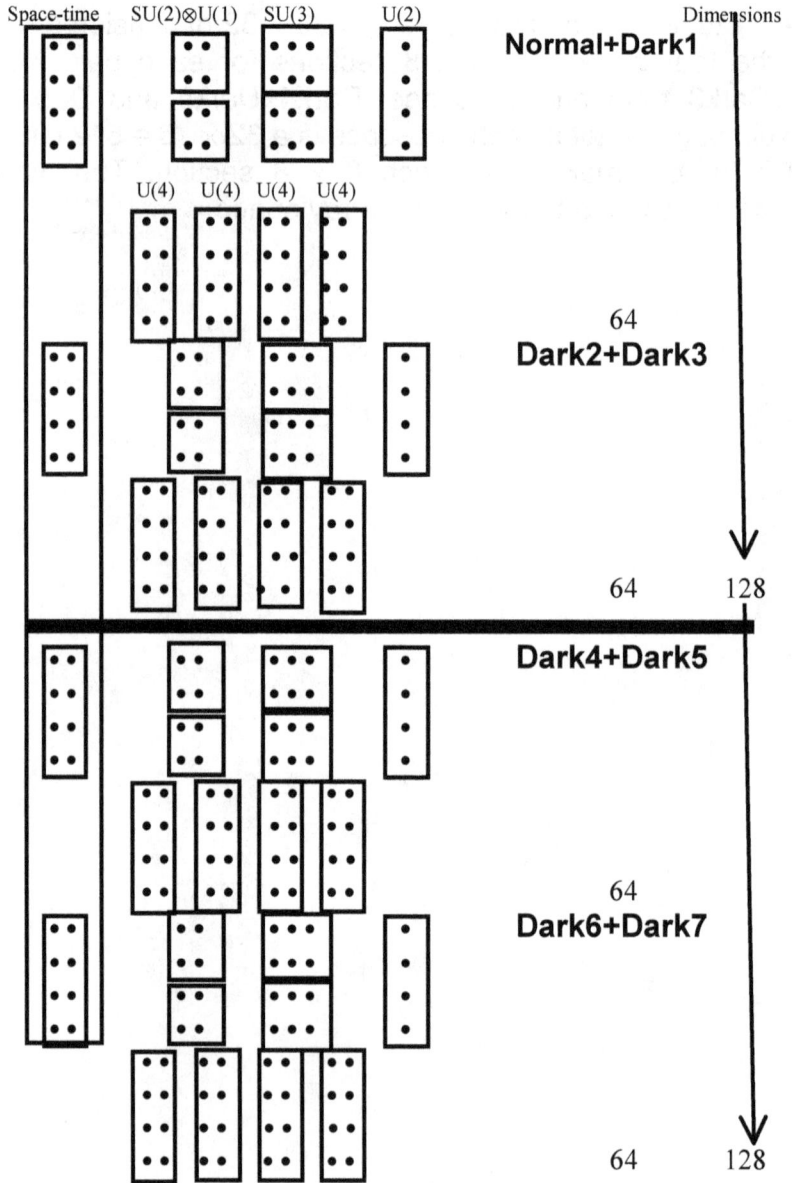

Figure 1.13s Partition of two of the eight rows in Fig. 1.12s for one of the four layers of UTMOST dimensions.

UTMOST Fermion Array

Normal	Dark1	Dark2	Dark3	Dark4	Dark5	Dark6	Dark7

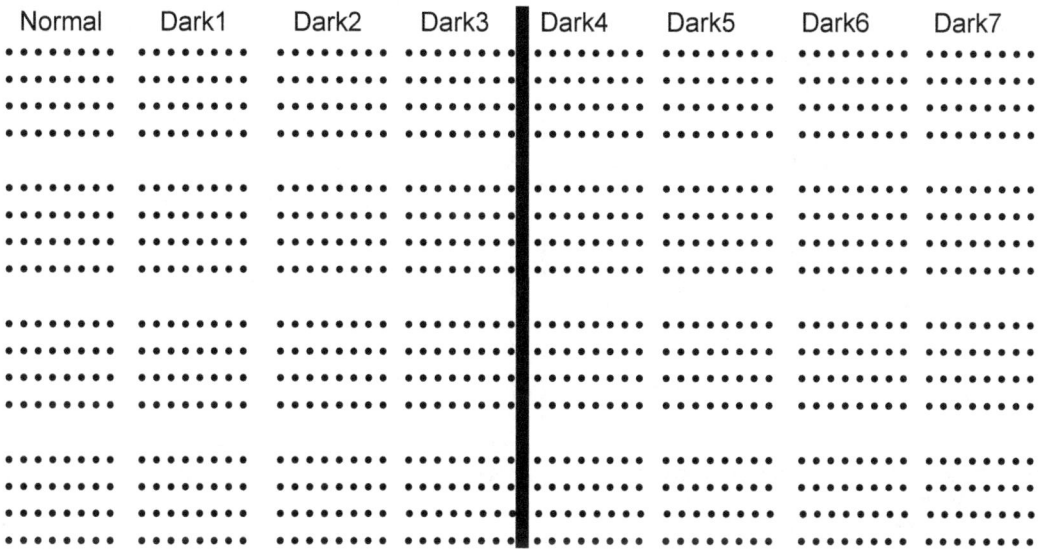

Figure 1.14s. Partition of spectrum of UTMOST fermions in a 16×64 format. Each fermion is represented by a •. Including each quark. Each set of eight •.'s represents a charged lepton, a neutral lepton, three up-type quarks, and three down-type quarks. There are eight sets of four species in four generations which are in turn in 4 layers. There are 512 fundamental fermions in each subspace taking account of quark triplets. Note: Quark singlets won't do; triplets are required.

	Normal		Dark1		Dark2		Dark3	
	e q-up **4**	v q-down **4**	e q-up **4**	v q-down **4**	e q-up **4**	v q-down **4**	e q-up **4**	v q-down **4**
4								
4								
4								
4								

	Dark4		Dark5		Dark6		Dark7	
4								
4								
4								
4								

Figure 1.15s. Partition of block form of the 32×32 UTMOST fermion array with each row corresponding to *half of a layer*. (Compare to Fig. 1.12s.) Thus $8 \times \frac{1}{2} = 4$ layers results. Each block contains four generations of fermions. The result is sixty-four 4×4 blocks. The label e q-up indicates a charged lepton – up-type quark pair, ν q-down indicates a neutral lepton – down-type quark pair, and so on.

The preceding figures clearly show that the partition of the UTMOST space into two 512 dimension parts gives two MOST subspaces.

6. Partition of MOST into QUeST Subspaces

A MOST space can be separated into two QUeST subspaces. They are described by the following figures. The QUeST subspaces can be further partitioned as shown in preceding chapters.

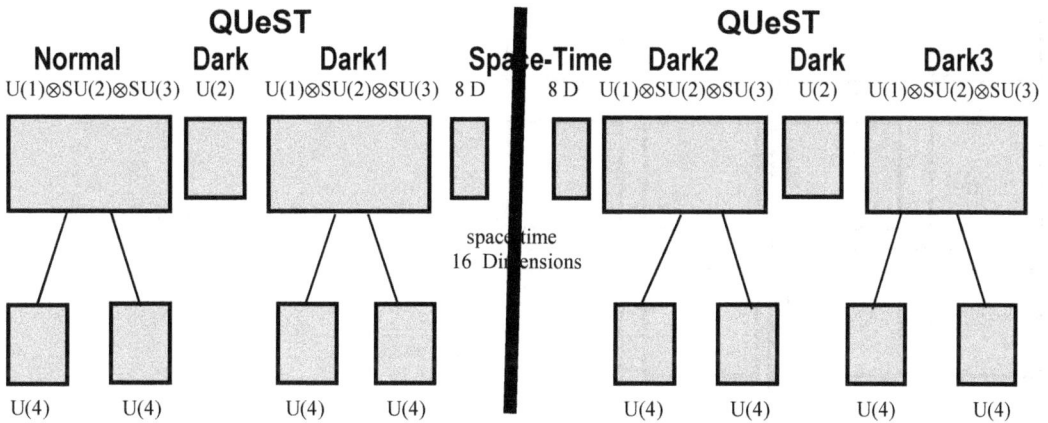

Figure 1.6s. The internal symmetry groups and space-time of *one* MOST layer split into two QUeST dimension arrays. The other three layers are similarly split.

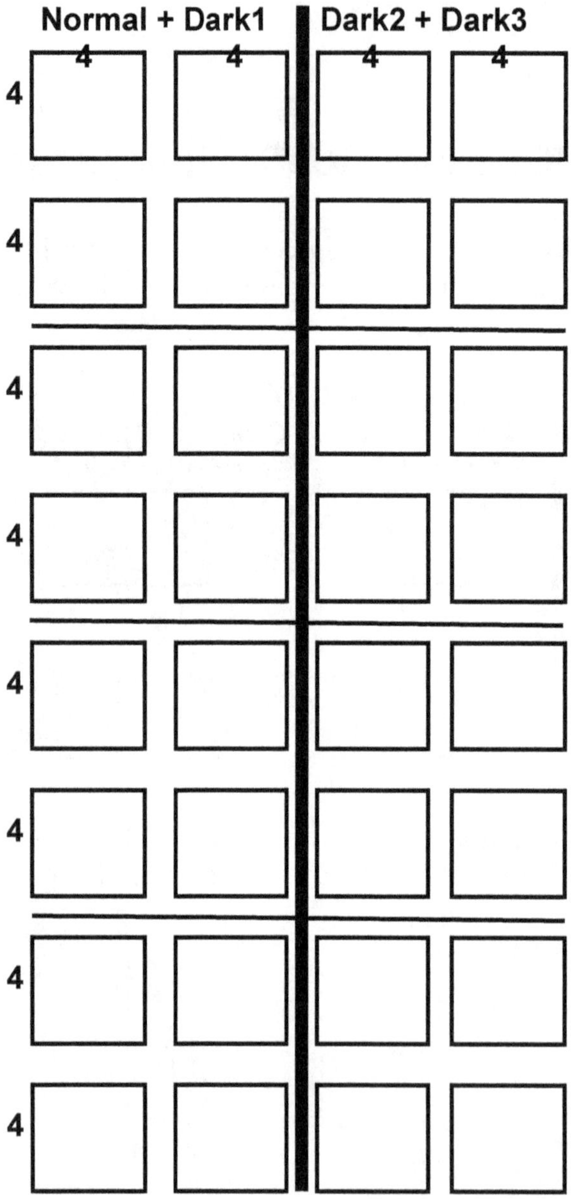
Normal + Dark1
Dark2 + Dark3
4
4
4
4
4
4
4
4
4
4
4
4

Figure 1.7bs. Four MOST layers (8 rows) split into two QUeST dimension arrays. 4×4 blocks within the four block, 8×8 sections for each pair of Normal and Dark sets of dimensions. They form the $16 \times 32 = 512$ MOST dimension array. See Fig. 1.2 for the makeup of each 8×8 section.

MOST Fermion Array:

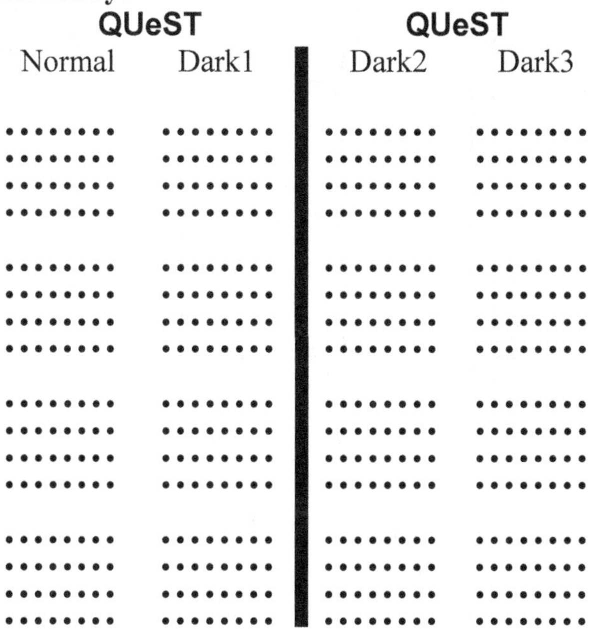

Figure 1.8s. Spectrum of the fermions of MOST split into two QUeST fermion arrays. Each fermion is represented by a •. Quark triplets are represented by a single •.

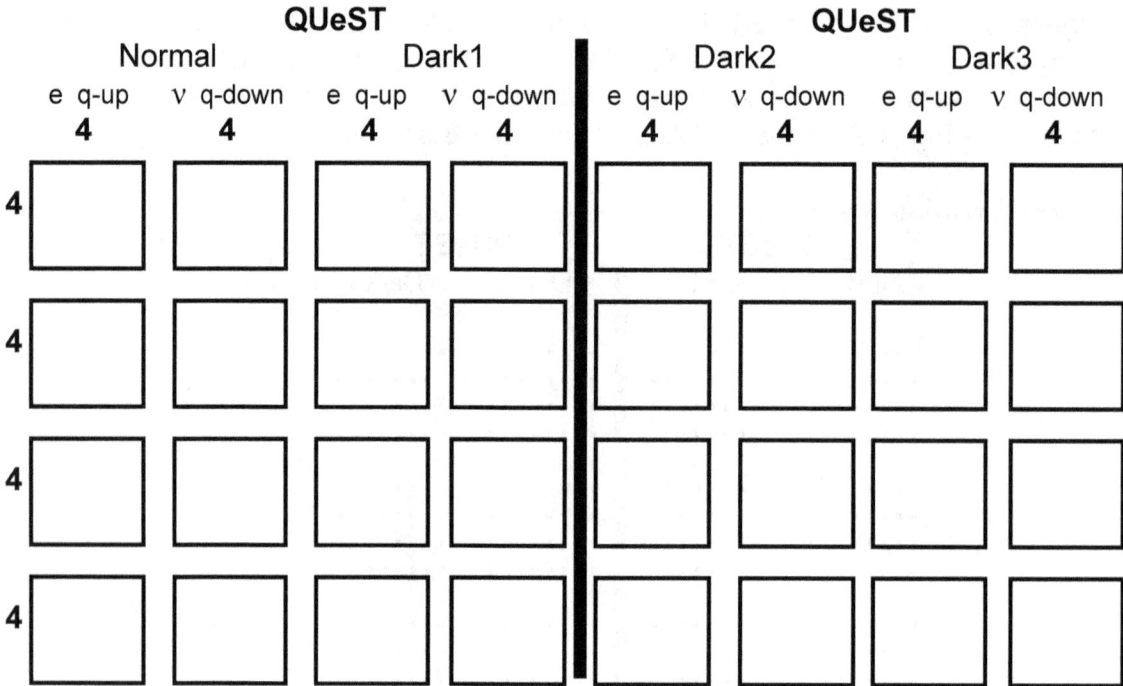

Figure 1.9s. Block form of MOST fermion array split into two QUeST parts. Each block row corresponding to one layer. Each block contains four generations of fermions. The result is 4 × 4 blocks. The fermion rows are reordered. The label e q-up indicates a charged lepton – up-type quark pair, ν q-down indicates a neutral lepton – down-type quark pair, and so on. *This form explains why generations come in fours.*

We have seen that MOST can be partitioned into two QUeST subspaces. As shown earlier each QUeST subspace can be partitioned as well.

7. Summary of QUeST and UTMOST Subspaces

7.1 UTMOST Subspaces

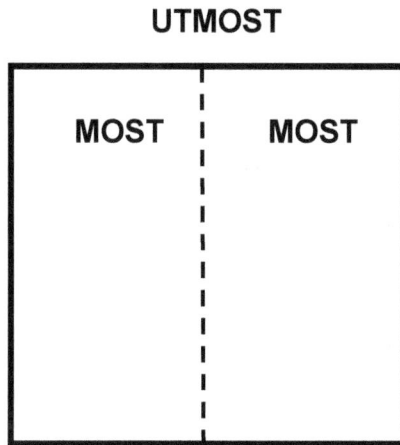

UTMOST

Figure 7.1. UTMOST division into two MOST subspaces.

7.2 MOST Subspaces

MOST

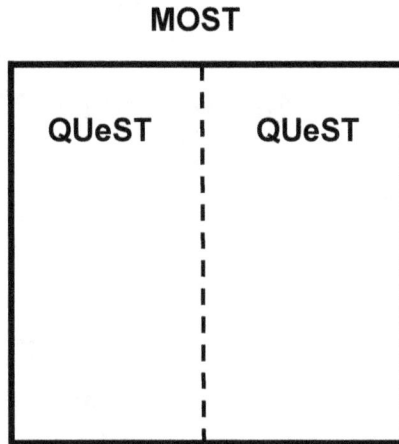

Figure 7.2. MOST division into two QUeST subspaces.

7.3 QUeST Subspaces: REAL and IMAGINARY

QUeST

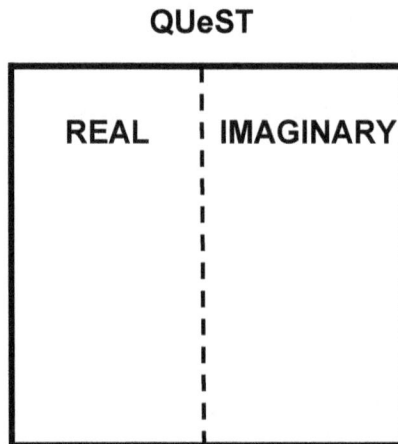

Figure 7.3. QUeST division into a REAL and IMAGINARY subspace. See chapter 4.

7.4 "Complete" UTMOST Subspaces

UTMOST

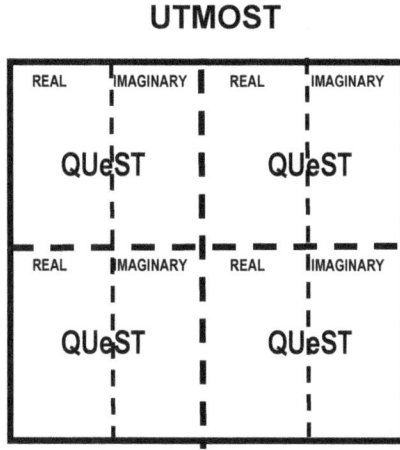

Figure 7.4. UTMOST division into two MOST subspaces, then two four QUeST subspaces, and finally into four REAL and four IMAGINARY "subsubspaces.".

8. Is There a Yet Deeper Stratum Below BQUeST and BMOST?

8.1 Concept of a Dimension

There are many definitions of a dimension. They all seem to be operational notions that define methods of calculating dimensions. One example of an operational definition is that of a fractal dimension.

8.2 Fractal Dimensions

The determination of fractal dimensions is operational. A fractal dimension specifies the scaling of a line segment (and other geometrical figures). It determines the change in the length of a line as the scale distance decreases. Typically the length of a jagged line increases as the scale decreases. Thus the fractal dimension D can be determined "experimentally". D is greater than one in the case in nature usually studied. The case D =1 corresponds to scale invariance. A straight line is a case of D = 1.

The case D < 1 does not seem to have been studied for lack of known "realistic" examples. In this case the length of a line *decreases* as the scale is decreased. In the limit as the scale goes to zero, the length of the line goes to zero. It appears possible to make a mathematical example using a broken straight line with a dense set of infinitesimal gaps.

8.3 Deeper Stratum?

Dimension seems to be best regarded as a primitive rather than as a defined concept since they do not appear to be definable in terms of deeper primitives. In view of this observation it appears the one dimension basis of BQUeST and BMOST would appear to be the lowest stratum of Physical Reality.

Beneath the Quaternion Universe:

UST, QUeST, BQUeST, UTMOST, BMOST

Beneath the Quaternion Universe:
UST, QUeST, BQUeST, UTMOST, BMOST

Stephen Blaha Ph. D.
Blaha Research

Derivation of QUeST Quaternion Dimensions from One Dimension BQUeST
BQUeST: 1 Dimension, 1 Fermion Implies QUeST
Derivation of MOST Octonion Dimensions from BMOST
BMOST: 1 Dimension, 1 Fermion Implies MOST and UTMOST
The Source of Hypercomplex Coordinates
"Hidden" U(8) Symmetry Basis of Internal Symmetries and Fermions
Block Structure of QUeST and UTMOST Dimensions and Fermions
One Dimension–One Fermion Proto-Universes and the Megaverse
Factorization of UTMOST into TWO MOSTs
Separated UTMOST Space-Times

Pingree-Hill Publishing
MMXX

ISBN: 978-1-7345834-8-9

Rev. 00/00/01   August 3, 2020

INTRODUCTION

Experimental High Energy Physics has become increasingly constrained by the cost of new high energy accelerators. Theoretical High Energy Physics has had the benefit of an enormous amount of data accumulated in the past century. Based on this data the Standard Model of Elementary Particles was developed. In recent years the author has constructed a theory of elementary particles called the Unified SuperStandard Theory (UST) that contains the Standard Model and, based on a set of axioms, develops an extended theory using an approach similar to Euclid's Geometry. This approach generated a larger internal symmetry group that the author recently found (January, 2020) was *exactly present in a 32 complex quaternion dimension theory* (256 dimensions) that he called QUeST. QUeST leads directly to UST upon restriction to real-valued 3+1 dimension coordinates. QUeST combines internal symmetries and space-time.

In a series of books in 2020 the author further explored the features of QUeST such as fermion-dimension duality. He also developed a Megaverse version called MOST (Megaverse Octonion SuperStandard Theory) with a 7+1 dimension space-time within a 512 dimension space. MOST is a 32 complex octonion theory that combines internal symmetries and space-time.

In this book we describe a new deep foundation, called BQUeST, for QUeST (and thereby UST). This foundation initially specifies a single dimension, and then derives 256 dimension QUeST with a new suggestive alignment of internal symmetries. The single dimension space requires a 7+1 dimension space-time, in which resides a fermion. The 7+1 dimension space is the same as MOST space-time. Thus a single fermion of sufficient energy in MOST's Megaverse can be the seed of a universe such as ours.

Subsequently we consider the foundation of Megaverse MOST, and a generalization of MOST called UTMOST. We found that UTMOST could be derived from a one dimension theory with one fermion called BMOST. The fermion resided in a separate 10 dimension space. Thus one could view the Megaverse as generated from a one dimension space.

The basis of QUeST and UTMOST in one dimension spaces with one primal fermion is reminiscent of the Pre-Socratic Philosophers search for the one element, of which all things were made. Our searches led to sole fermions and dimensions, from which all things follow.

There are several good books for these theories. UST is described in Blaha (2020c); QUeST and MOST are described in Blaha (2020g), which also introduces BQUeST.

1. Some Features of the Unified SuperStandard Theory (UST) and the Quaternion Unified SuperStandard Theory (QUeST)

Having successfully based the Unified SuperStandard Theory (UST) on the Quaternion Unified SuperStandard Theory (QUeST) with a remarkable match between the internal symmetries and space-time symmetry of both theories, we now address the issue of a deeper basis for QUeST. In this chapter we briefly outline some features of UST and QUeST.

1.1 Axioms for the Unified SuperStandard Theories

Our derivation of UST was based on Complex General Relativity, and Quantum Field Theory using Two Tier coordinates and PseudoQuantum fields.

It is a derivation in the manner of Euclid's Geometry that starts from the axioms and noting that a host of conservation laws follow from free quantum field theory, develops an extended set of internal symmetries including Generation groups and Layer groups. Based on an analogy with Complex Special Relativity subgroups, $U(1) \otimes SU(2) \otimes SU(3)$ appears. The resulting internal symmetries are

$$[SU(2) \otimes U(1) \otimes SU(3) \otimes SU(2) \otimes U(1) \otimes SU(3) \otimes U(4)^4 \otimes U(2)]^4$$

which includes a set of four $U(2)$ groups that map between the Normal and Dark fermion sectors. Almost all internal symmetries are badly broken.

We begin by listing a set of Axioms for the combined Unified SuperStandard Theories for our $3 + 1$ dimension universe, for the $3 + 1$ complex quaternion QUeST universe, and for the $7 + 1$ octonion MOST Megaverse.

AXIOMS

1. A complex quaternion space is the basic space of our universe. Complex octonion space is the basic space of the Megaverse.

2. Physical processes can execute in parallel.

3. Matter and energy are particulate.

4. Space--times are locally Lorentzian.

5. All calculations are finite.

6. Particle theory can be defined in any curved space-time.

7. Each particle has a wave function determined by a functional inner product defining the particle state. The functionals form a set without a distance measure.

Figure 1.1. Axioms of UST and QUeST.

1.2 Immediate Implications of the Axioms

In this section we describe some of the implications of each of the axioms.

1. Complex quaternion space is the basic space of our universe. Complex octonion space is the basic space of the Megaverse.

The factorization into a space-time and an internal symmetry space results from symmetry breaking. It appears to be related to a breakdown of the vacuum.

2. Physical processes can execute in parallel.

Physical processes are known to be able to execute in parallel at any distance of separation. As Fant has shown parallel execution requires a minimal number of dimensions: 4. Consequently the dimension of space-time must be 4 or greater. The biquaternion space-time of QUeST is 4-dimensional allowing parallel process execution.

The bioctonion space-time of MOST is 8-dimensional and also allows parallel process execution. The choice of eight dimensions is natural since it allows 4-dimensional universes within it. It also has a form that allows a clean formulation. Lastly, as will be seen later, it conforms to the pattern of interplay between Lorentz symmetry and internal symmetry found in the Unified SuperStandard Theory. This axiom leads to a view of the origin of the dimensions.

3. Matter and energy are particulate.

The most direct method of specifying a theory of matter and energy is through the Use of Quantum Field Theory. Thus Quantum Field Theory is implied.

4. Complex Space-times are locally Lorentzian.

A locally complex Lorentzian space-time leads to Complex General Relativity. In flat space-time Complex General Relativity becomes Complex Lorentz group. (In point of fact the Complex Poincaré group follows.)

5. All calculations are finite.

Given the need for Quantum Field Theory it becomes necessary to find a formulation that yields finite values for calculations in perturbation theory. The only approach that eliminates high energy divergences, and yet preserves the results found in perturbation theory calculations that agree with (primarily QED) experiments, is Two-Tier Quantum Field Theory. This is discussed in detail in earlier books starting in 2002. Thus only our Two-Tier formalism satisfies this axiom.

6. Particle theory can be defined in any curved space-time.

In the 1970s we developed a formalism that allows the definition of particle states in any space-time in such a way that its physical content is preserved when transformed to any coordinate system.[12] This PseudoQuantum Quantum Field Theory satisfies this axiom.

7. Each particle has a wave function determined by a functional inner product defining the particle state. The functionals form a set without a distance measure.

This axiom is satisfied by our formulation of quantum functionals in Blaha (2019f) and earlier books. Our formulation eliminates the superficial violation of the Theory of Relativity by "spooky" quantum entangled processes with parts separated by a physically "large" distance.

The seven axioms imply the Unified SuperStandard Theory and its deeper hypercomplex formulations: QUeST and MOST.

1.3 Some UST Features

UST symmetries are directly derivable from QUeST. We will therefore begin with a QUeST summary and then proceed to describe features that are common to both UST and QUeST.

QUeST was formulated in 32 dimension complex quaternion space with a total number of 256 dimensions. The number of fundamental fermions was found to number 256. Figs. 1.2a and 1.2b display the symmetries of UST and QUeST.

Fig. 1.3 lists the fermion periodic table in UST and QUeST including the currently known fermions. It also shows the interactions between the fermions. Notice all fermions are connected in the sense that interactions exist to transform any fermion into any other fermion. *The interconnections of the fermions suggest that there is a commonality between them, which the deeper basis proposed in this chapter supports.*

---

[12] S. Blaha, Il Nuovo Cimento **49A**, 35 (1979).

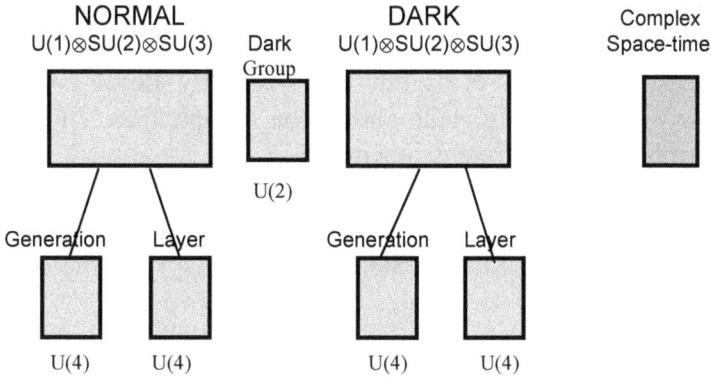

Figure 1.2a. Schematic of the internal symmetry groups for *one layer* including 4 complex dimension space-time. The two large blocks are each 5 dimension complex coordinate representations of SU(2)⊗U(1)⊗SU(3). The U(2) group, called the Dark group, supports transformations (rotations) between normal and Dark matter.

The symmetries in Fig. 1.2a include the Standard Model symmetries. They also contain Generation groups and Layer groups for both Normal and Dark matter.

　　The Generation groups mix the fermion generations of Normal and Dark sectors within each layer. The lines on the left side of Fig. 1.3 displays Generation group mixing in each layer.

　　Layer groups mix fermions in all four layers for each of the four generations individually. (See right side of Fig. 1.3.) In four layer UST and QUeST there are eight Layer groups: two Layer groups for Normal and Dark sectors for each of the four generations. Although the eight Layer groups are listed in Fig. 1.2b as two Layer groups in each layer of UST and QUeST, Fig. 1.3 shows the mixing takes place between layers, generation by generation.

　　The eight Layer groups and eight Generation groups in Figs. 1.2a and 1.2b are present in UST and QUeST as well as being implied by BQUeST. These groups are all U(4) groups because each U(4) group has its own set of four conserved (in free field approximation) particle numbers.

The U(2) Dark group mixes between Normal and Dark fermions, individual fermion by fermion, as shown in Fig. 1.3. This group surfaced in QUeST. It was not originally in UST. It is now added.

Fig. 1.2a shows the set of internal symmetry fields and complex space-time coordinates *for one layer*. Fig. 1.2b shows the complete set of four layers of QUeST dimensions numbering 256 dimensions in total.

QUeST and UST have the internal symmetry:

$$[SU(2) \otimes U(1) \otimes SU(3) \otimes SU(2) \otimes U(1) \otimes SU(3) \otimes U(4)^4 \otimes U(2)]^4 \qquad (1.1)$$

In addition they have 4 complex space-time coordinates in the case of UST and 4 complex quaternion space-time coordinates in the case of QUeST.

The allocation of dimensions to fundamental representations of internal symmetry groups uses the rule:

U(2) requires 4 real dimensions
U(1)⊗SU(2) requires 4 real dimensions
SU(3) requires 6 real dimensions
U(4) requires 8 real dimensions

for *real dimensions* with *real-valued* coordinates. The internal symmetry groups in UST and QUeST are:

SU(3)	Strong Interaction group
U(1)⊗SU(2)	ElectroWeak group
U(4)	Generation and Layer groups
U(2)	Dark group

1.4 Fermion-Dimension Duality

The fermions and dimensions of QUeST number 256 each. They display a duality, which was described in Blaha (2020g). We will consider the implications of this duality in more detail in subsequent chapters.

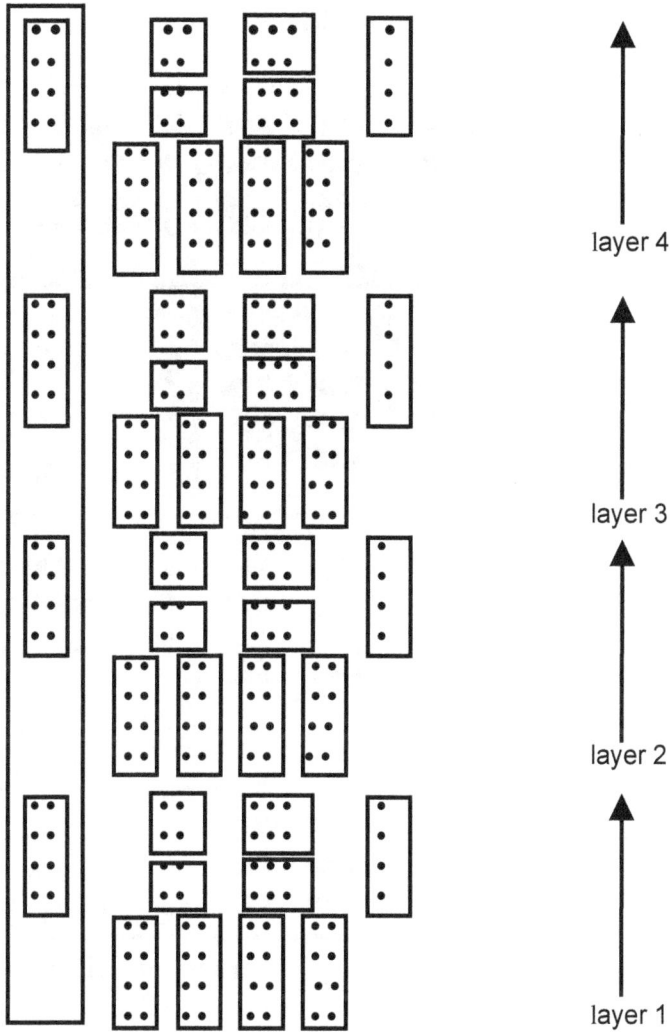

Figure 1.2b. Four layer UST (and QUeST) internal symmetry groups and space-time diagram for 32 dimension complex quaternion space. Note the left column of composite blocks combine to give a 4 dimension complex quaternion space-time.

The Fermion Periodic Table

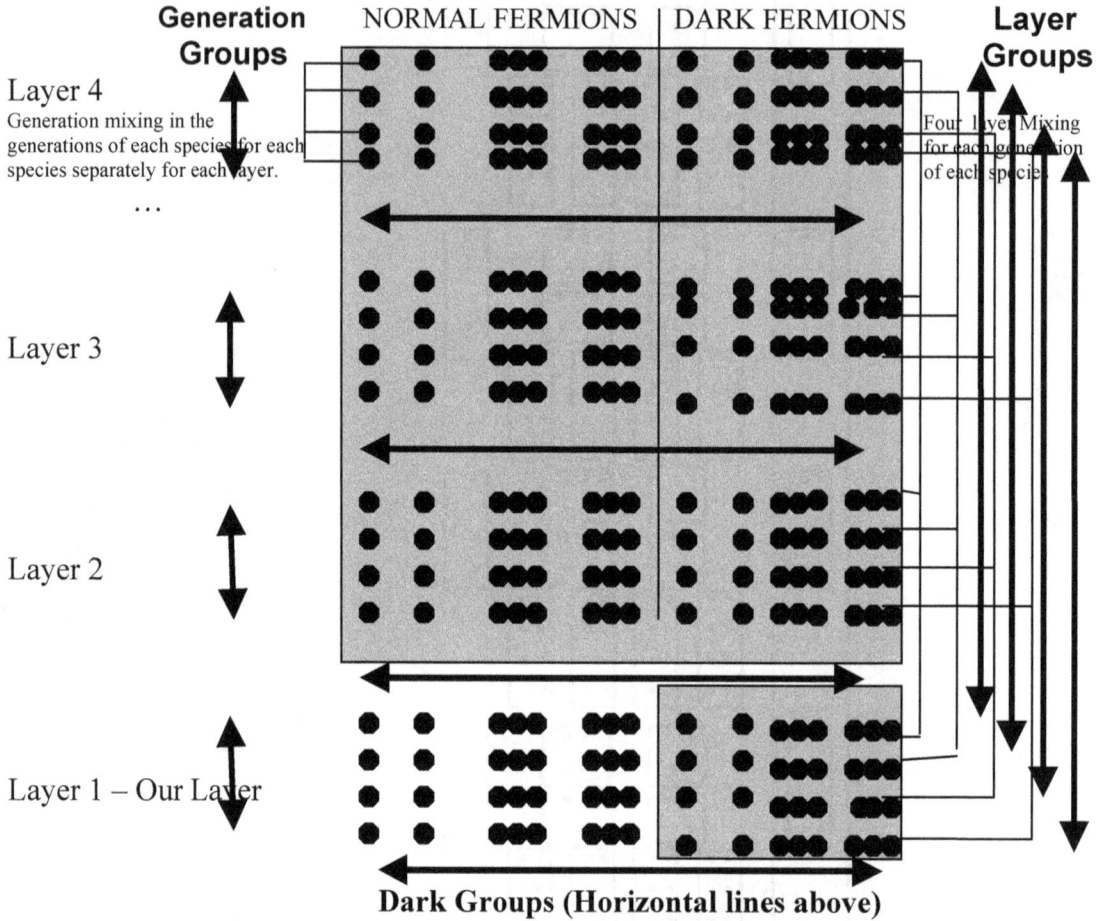

Figure 1.3. Fermion particle spectrum and partial example of pattern of mass mixing of the Generation, Layer, and Dark grroups. Unshaded parts are the known fermions including an additional, as yet not found, 4th generation. Shaded parts are the truly Dark fermions *and* the, as yet unknown, Normal

fermions. The lines on the left side (only shown for one layer) display the Generation mixing within each layer's species. The Generation mixing applies within each layer using a separate Generation group for each layer. The lines on the right side show Layer group mixing with the mixing amongst all four layers for each of the four generations individually. There are four Layer groups. The Dark groups mixing between normal and Dark fermions are shown in the center as horizontal lines. There are 256 fundamental fermions counting quarks as triplets.

DIMENSIONS		FERMIONS			
Real	Imaginary	e	v	up-q	down-q

Layer 1

```
....  ....        .  .  ...  ...
....  ....        .  .  ...  ...
....  ....        .  .  ...  ...
....  ....
```

DARK

		e	v	up q	down q

```
....  ....        .  .  ...  ...
....  ....        .  .  ...  ...
....  ....        .  .  ...  ...
....  ....
```

Layer 2

```
....  ....        .  .  ...  ...
....  ....        .  .  ...  ...
....  ....        .  .  ...  ...
....  ....        .  .  ...  ...
....  ....        .  .  ...  ...
....  ....        .  .  ...  ...
....  ....        .  .  ...  ...
```

Layer 3

```
....  ....        .  .  ...  ...
....  ....        .  .  ...  ...
....  ....        .  .  ...  ...
....  ....        .  .  ...  ...
....  ....        .  .  ...  ...
....  ....        .  .  ...  ...
....  ....        .  .  ...  ...
```

Layer 4

```
....  ....        .  .  ...  ...
....  ....        .  .  ...  ...
....  ....        .  .  ...  ...
....  ....        .  .  ...  ...
....  ....        .  .  ...  ...
....  ....        .  .  ...  ...
....  ....        .  .  ...  ...
```

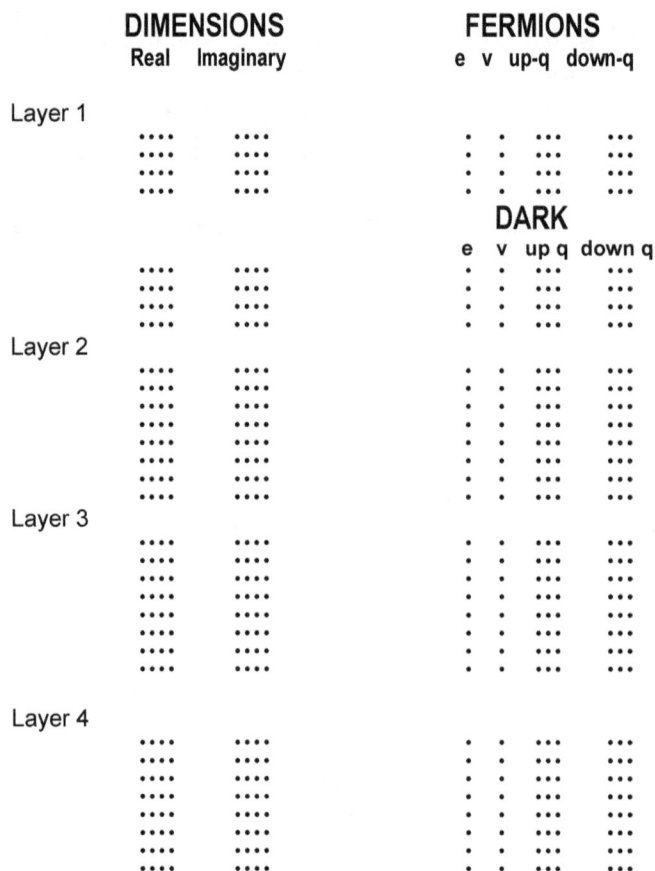

Figure 1.4. Fundamental fermions have a 1:1 correspondence with QUeST dimensions. Note the number of dimensions in each row is 8 – the number of dimensions in a complex quaternion. Correspondingly the number of fermions in each row is 8 – a suggestive similarity. Each layer has four normal fermion generations and four Dark fermion generations. Each dot (pebble) represents a dimension in the left part and a fermion in the right part.

2. The Structure of Thirty-Two Complex Quaternion QUeST Space

In this chapter we construct a deeper basis for QUeST in a smaller dimension space. Our goal is to free QUeST from its 256 dimensions and 256 fermions by seeking a more fundamental grounding.

2.1 A 16 × 16 Form of the QUeST Dimension Array

The dimension array of 32 complex quaternion dimension QUeST is displayed in Fig. 2.1.

Figure 2.1. The 256 dimensions of QUeST.

We simply reorder the rows and columns to make a 16 × 16 array which will be more convenient for our derivation of a more fundamental basis for QUeST that we will call BQUeST (pronounced bee-quest). No change in the dimensions is made. (BQUeST could also be viewed as residing in a 16 complex octonion dimension space for the purpose of dimension counting.)

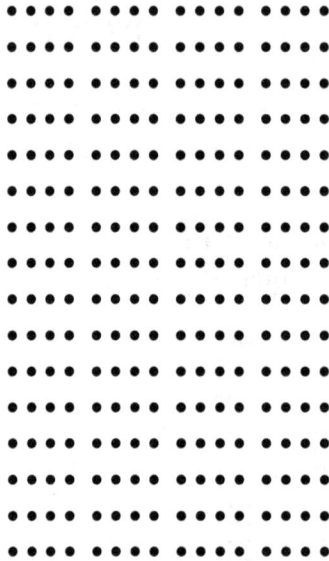

Figure 2.2. 16 × 16 array of QUeST dimensions.

2.2 QUeST Dimensions Built from One Dimension

We begin by assuming a one dimension space where the dimension has an associated dimension functional with arguments of spin and momentum but not explicitly internal symmetries. We call it the *Q-basis*.

To obtain a 16 × 16 array of dimensions we assume the Q-basis dimension functional resides in a separate eight dimension space. The spin of the functional then has 16 components.

Then we assume a single fermion, which we call a *Q-fermion*, with two fermion fields,[13] each with 16-spinors. Treating the spinor components as independent[14] we can form a 16×16 array of functionals

$$D_{ij} = f(p_1, s_{1i})f((p_2, s_{2j}) \qquad (2.1)$$

which we take to be the functionals of QUeST. The QUeST dimension array is then

$$D_{Qij} = D_{ij} \qquad (2.2)$$

with composite functionals

$$f_Q(p_1, p_2) \qquad (2.3)$$

Thus we obtain the 16×16 QUeST dimension array.

2.3 Eight Dimension Space of the Megaverse

The price of generating the QUeST dimensions from a one dimension space is the introduction of an eight dimension space for new fundamental fields, Q-fermions. At first glance this situation is problematic. However we note that our Megaverse MOST formulation has a 7+1 complex octonion space-time.

2.3.1 Eight Dimension Space is MOST Space

Thus we may assume, provisionally that the eight dimension space is the 7+1 MOST space-time of the Megaverse.

---

[13] Blaha's PseudoQuantum formulation of Quantum Field theory has two fields representing each fermion and boson particle. It has the advantages of allowing a canonical formulation of quantum field theory with higher order differential equations such as those needed for quark confieement, and of allowing quantum field theory formulations in *any* space-time. See S. Blaha, Phys Rev **D10**, 4268 (1974); _____, **D11**, 2921 (1974) for PseudoQuantum quark confinement with an r potential. See S. Blaha, Phys. Rev **D17**, 994 (1978), _____, IL Nuovo Cimento **49A**, 35 (1979), _____, IL Nuovo Cimento **49A**, 113 (1979) for PseudoQuantum formulation in arbitrary space-times using Bogoliubov transformations..

[14] The spinor components of each fermion field is independent of the others field's spinor components.

2.3.2 Origin of Universe

We can then envision the possibility that our universe started as 1-dimensional Q-fermion and then "acquired" the 256 dimensions of QUeST and 256 fundamental fermions to form a space that includes our 3+1 space-time at the Big Bang point.

Consider the universe as beginning as a Q-fermion[15] in the Megaverse and dynamically growing to be the current universe. This view is supported by the derivation of the universe growth rate (Hubble rate) near the Big Bang as a particle vacuum polarization effect. See Blaha (2019e).

The consequence is a form of unification of our universe and the Megaverse.

2.4 Iterative Generation of QUeST Dimensions

We can establish the 256 dimension QUeST array in a two-step process. First we define a Q-fermion with a functional that has 4-spinor components in a 3+1 space-time. Each component of the off-shell Dirac field functional is independent. We then define the two Q-fermion functionals

$$f_1(p, s) \quad \text{and} \quad f_2(p, s) \tag{2.4}$$

with each functional, $f_1(p, s)$ and $f_2(p, s)$, having four spinor components. Thus their product forms a 4×4 array.

$$F(i, j) = f_1(p, s_i) \, f_2(p, s_j) \tag{2.5}$$

The indices i, j = 1,2, 3, 4 will subsequently label 4×4 blocks of functionals.

Next, this 4×4 dimension (block) array generates the QUeST 16×16 array of dimensions by viewing each generated functional $F(i, j)$ as describing a block of 4×4 = 16 dimensions. This is done by treating each $F(i, j)$ block as the product of two Dirac functionals of the same form as eq. 2.4 with 4-spinor components.

---

[15] The origin of the universe in a spin ½ particle might imply a "spinning" of the universe that is not known at present. A zero spin universe might be generated from a spin zero dimension field particle corresponding to the one dimension of the Q-basis.

Mapping to dimensions gives the QUeST array of 256 dimensions. This map to a 16×16 dimension array is diagrammed in Fig. 2.4.

2.4.1 New View of Block Structure of QUeST array

The intermediate form of the 4×4 dimension array suggests a new form of structuring the 256 dimension QUeST array as composed of 16 blocks containing 4×4 = 16 dimensions.

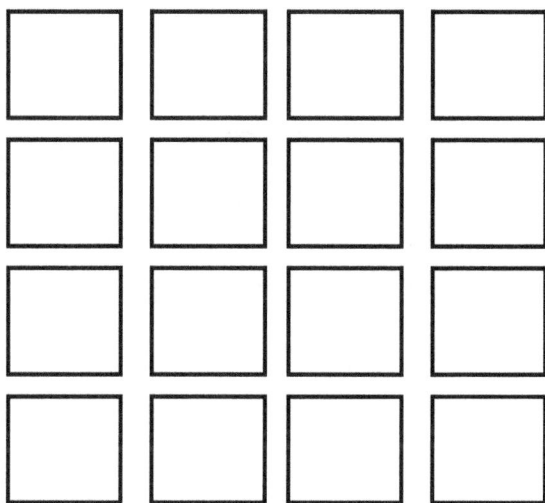

Figure 2.3. Block form of 16 × 16 QUeST dimension array with each row corresponding to one layer.

The block form of Fig. 2.3 has significant implications for the internal symmetry structure of the QUeST array and for the structure of the 256 set of fermions.

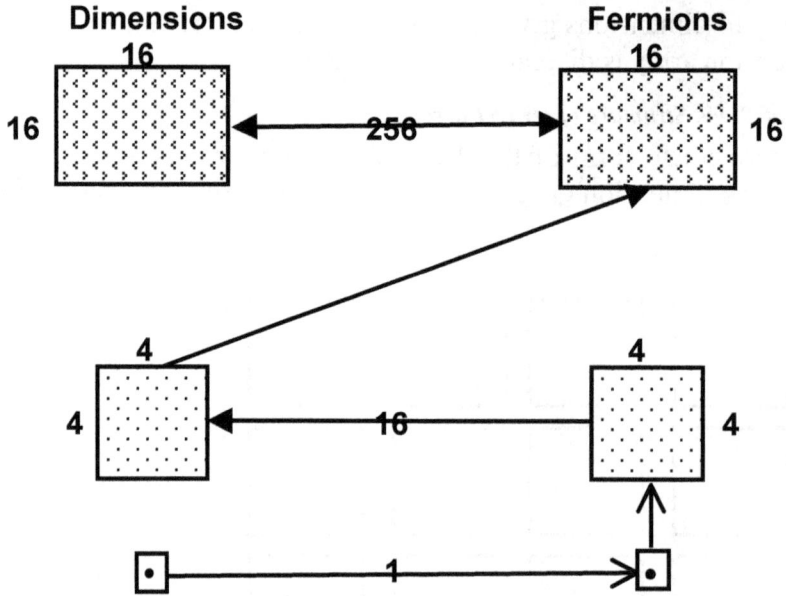

Figure 2.4. Diagram of the ascent to the QUeST array.

2.5 Known Symmetry Group Structure

Fig. 2.5 displays the form of the QUeST dimension array with subgroups indicated by boxed dimensions for *one layer* of dimensions.

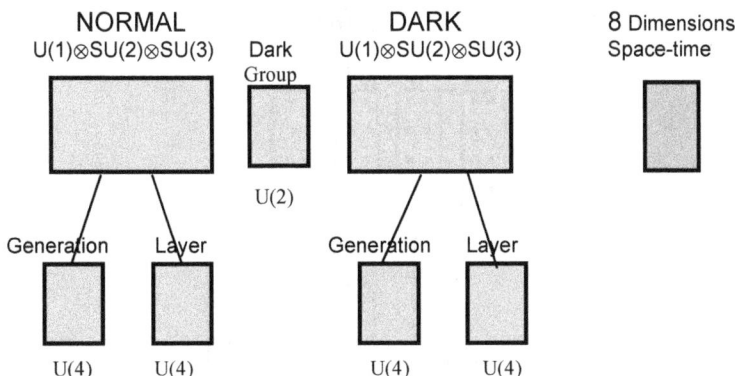

Figure 2.5. Schematic of the internal symmetry groups for one QUeST layer. Including 8 dimensions for space-time. The two large blocks are each 5 complex dimension representations of SU(2)⊗U(1)⊗SU(3). The U(2) group supports transformations (rotations) between normal and Dark matter.

Fig. 2.5 shows the set of internal symmetry fields and complex space-time coordinates *for one layer*. Fig. 2.6 shows the complete set of four layers of QUeST dimensions numbering 256 dimensions in total.

QUeST and UST have the internal symmetry:

$$[SU(2) \otimes U(1) \otimes SU(3) \otimes SU(2) \otimes U(1) \otimes SU(3) \otimes U(4)^4 \otimes U(2)]^4 \qquad (2.6)$$

In addition they have 4 complex space-time coordinates in the case of UST and 4 complex quaternion space-time coordinates in the case of QUeST.

The allocation of dimensions to fundamental representations of internal symmetry groups uses the rule:

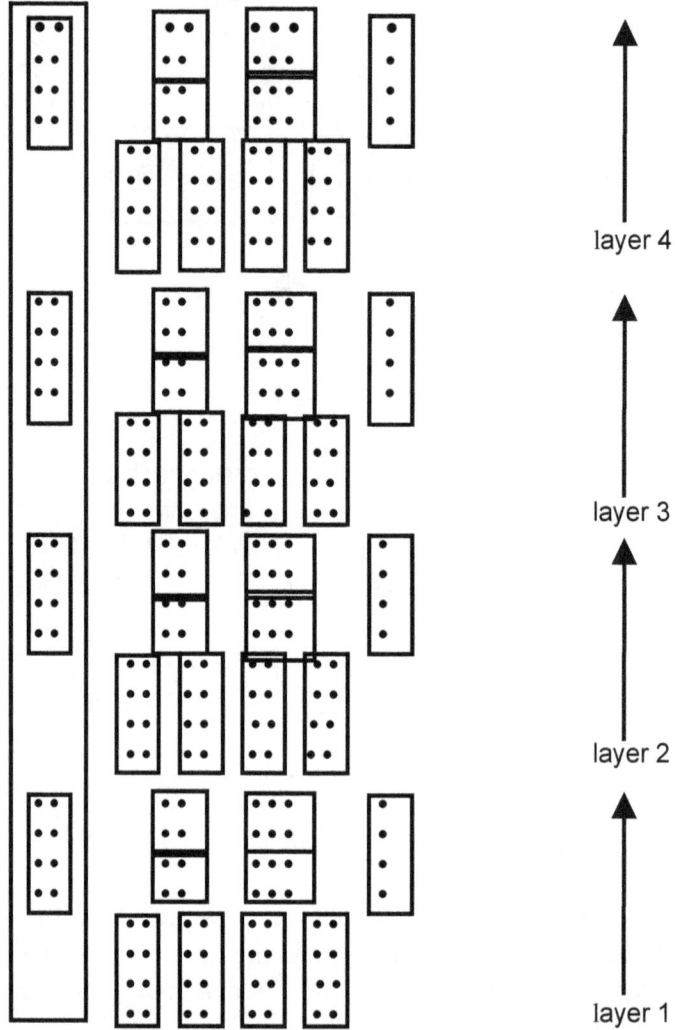

Figure 2.6. Four layer QUeST internal symmetry groups and space-time diagram for 32 dimension complex quaternion space. Note the left column of composite blocks combine to specify a 4 dimension complex quaternion space-time.

U(2) requires 4 real dimensions
U(1)⊗SU(2) requires 4 real dimensions
SU(3) requires 6 real dimensions
U(4) requires 8 real dimensions

where the dimensions have *real-valued* coordinates and are called *real dimensions*.

The eight Layer groups and eight Generation groups in Figs. 2.5 and 2.6 are present in both UST and QUeST as well as being implied by BQUeST.

2.5.1 Generation and Layer groups of UST, QUeST and BQUeST

The Generation groups mix the fermion generations of normal and Dark sectors of each layer The lines on the left side of Fig. 1.3 displays Generation group mixing in each layer.

Layer groups mix fermions in all four layers for each of the four generations individually. (See right side of Fig. 1.3.) There are eight Layer groups: two Layer groups for Normal and Dark sectors for each generation.

The Dark groups mix between normal and Dark fermions, fermion by fermion, as shown in Fig. 1.3.

2.6 QUeST and UST Dimension Restructure

We now will restructure the dimension array of Fig. 2.5 from the perspective of BQUeST. We will find the BQUeST dimension array breaks up into sixteen 16 dimension subblocks due to the 16-spinor substructure of BQUeST in Fig. 2.3. When this new subdivision is implemented for the array of 256 fermions of QUeST some remarkable conclusions follow.

Fig. 2.7 is the restructured (for one layer) Fig. 2.5 with the dimensions in blocks of 16 dimensions.

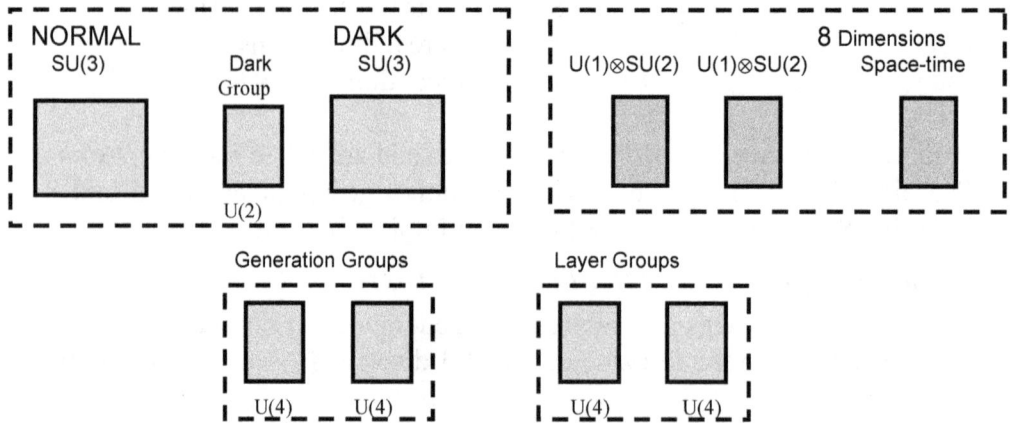

Figure 2.7. Schematic of the internal symmetry groups for one QUeST layer. Including 8 dimensions of space-time. Each block (regardless of size) contains 16 dimensions. The four blocks total to 64 dimensions. The U(2) group supports transformations (rotations) between Normal and Dark matter.

The four layers of QUeST, UST, and BQUeST are four copies of Fig. 2.7. Each of the layers has its own set of internal symmetry interactions. The four copies of the 8 space-time dimensions lead to a 3+1 complex quaternion space-time. The Layer groups also straddle the four layer copies as described above in section 2.5.1.

2.7 Some Implications of the New Dimension Array Structure

The 256 QUeST (and BQUeST) dimensions imply an initial fundamental representation of U(128). The block structure of Fig. 2.7 and the four layer QUeST structure suggest that there are two breakdowns: one breakdown to $U(8)^{16}$, and a "fine structure-like" breakdown to the internal symmetry groups of Fig. 2.6.

The subdivision into blocks of 16 dimensions has a number of implications:

1. Suggests a set of subgroups that may be relevant for symmetry breaking. The combination of $U(1)\otimes SU(2)$ groups with space-time (before further breakdown)

was presaged by a study of the relation of ElectroWeak symmetry to complex space-time in Blaha (2020c) and in earlier books by the author.

2. The combination of the SU(3) groups and the U(2) Dark group seems quite natural at the time before "fine structure" symmetry breakdown. (The Dark group rotates normal and Dark fermions with the same U(1)⊗SU(2)⊗SU(3) internal symmetry indices.)

3. The block with SU(3) and U(2) groups represents the combined internal symmetries before symmetry breakdown.

4. The Generation groups are combined as they should be before symmetry breaking. Similarly the Layer groups are combined.

5. The pattern of the blocks is in agreement with the assignment of QUeST dimensions to internal symmetry groups.

We conclude the block pattern of Fig. 2.7 and its four layer QUeST pattern are in accord with two symmetry breakdowns: a breakdown to $U(8)^{16}$, and a "fine structure-like" breakdown to the groups depicted in Fig. 2.6.

We have derived the complex quaternion dimensions of QUeST from BQUeST, and found a remarkable breakdown of U(128) to U(8) symmetries and a breakdown of the U(8) symmetries to those of UST.

3. The Structure of the QUeST Fermions

3.1 Fermion Structure

The structure of UST and QUeST fermions was displayed in Fig. 1.3. The possible $U(8)^{16}$ block structure described in sections 2.7 and 2.8 suggest a similar block structure for the fundamental fermions. Fig. 3.1 displays a set of sixteen 4×4 blocks with each block holding 16 fermions.

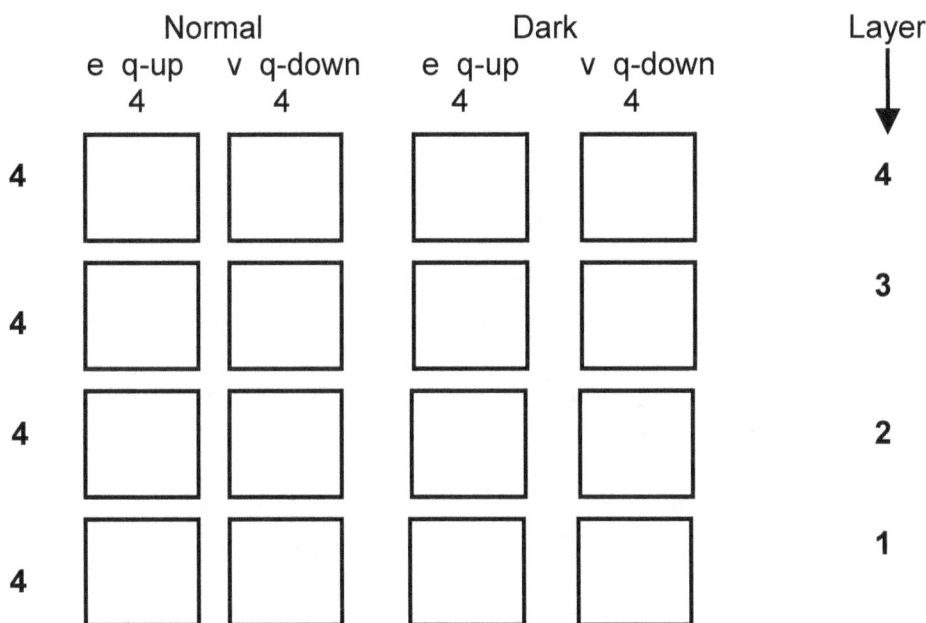

Figure 3.1. Block form of 16×16 QUeST fermion array with each block row corresponding to one layer. Each block contains four generations of fermions. The result is 4×4 blocks. The label e q-up indicates a charged lepton – up-type quark pair, v q-down indicates a neutral lepton – down-type quark pair, and so on.

The fermions are distributed among the blocks in a pattern similar to Fig. 1.3 with Normal fermions in the first two columns and Dark fermions in the 3rd and 4th columns. The fermions are reordered. The first and third columns have a charged lepton paired with an up-type quark. The second and fourth columns have a neutral lepton paired with a down-type quark.

3.1 Fermion Species and the Pairing of Fermions in Blocks

In our development of UST in Blaha (2020c) and earlier books we introduced the idea of four types of fermion species: charged lepton, neutral lepton, up-type quark, and down-type quark. We suggested that fermion species were a result of the application of four types of Complex Lorentz boosts. We found that it was natural to identify charged leptons as Dirac fermions, neutral leptons as tachyon fermions,[16] up-type quarks as Dirac-type fermions with complex 3-momentum, and down-type quarks[17] as tachyon fermions with complex 3-momentum.

The pairing of the leptons and quarks, which follows from the need to have 4 × 4 blocks, is based on pairing Dirac-like fermions in one type of block, and tachyon fermions in the other type of block in Fig. 3.1.

3.2 Reordered Fermion Structure Implications

A number of implications follow from the form of Fig. 3.1:

1. Each block has a quartet that is similar in form to a 3+1 coordinate system. This similarity raises the possibility of a Lorentz-like transformation group where a lepton plays the role of time and a quark triplet plays the role of spatial coordinates. A form of this possibility was considered in Blaha (2020c). It also raises the possibility of symmetry breaking along the lines of a Lorentz group.

2. The SU(2)⊗U(1) ElectroWeak group structure is clearly evident in the fermion blocks. Fig. 3.2 shows the charged generators of the group act between the first and second column blocks, and between the third and fourth column blocks (the

---

[16] Blaha (2020c) describes experimental data consistent with tachyon neutrinos.
[17] This possibility may explain 'x' cutoffs, and other modifications, needed to make parton calculations agree with deep inelastic scattering data.

horizontal arrows). The neutral charge group generators act vertically in each column (vertical arrows). ElectroWeak transformations can change ElectroWeak quantum numbers. They are diagonal in other particle internal quantum numbers.

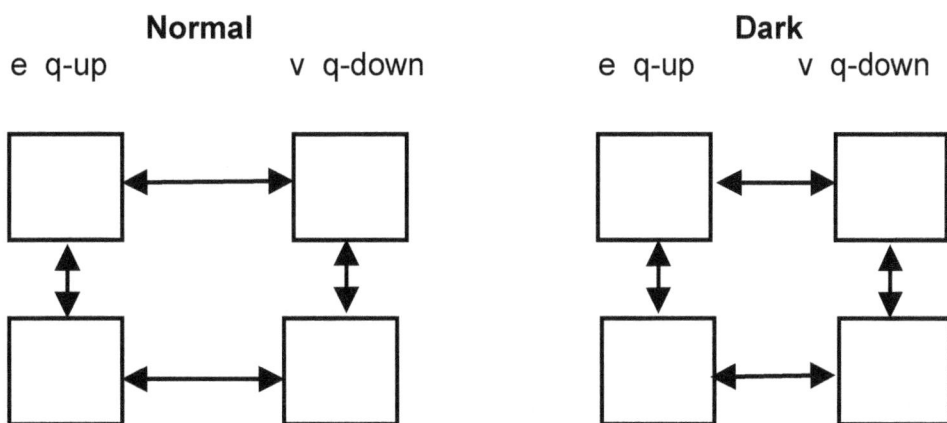

Figure 3.2. The effect of the ElectroWeak generator fields on the blocks of fermions.

4. One Dimension BQUeST and MOST (UTMOST)

In chapter 2 we described the one dimension precursor of QUeST that we called BQUeST where the initial 'B' denotes basis. We found BQUeST has one dimension and one fundamental fermion denoted the Q-fermion. The Q-fermion resides in an 8-dimension space and has a 16 component spinor. The spin assignment and thus the eight dimensions of the space-time were required for the generation of 256 dimension QUeST.

The relation of the eight dimension space-time to the 256 dimensions of QUeST was not exactly specified. In QUeST the 256 dimensions of the space contained a 3+1 complex quaternion dimension space-time subspace. The fundamental fermions resided in this space-time.

In the case of BQUeST the one dimension space has a matching fundamental fermion, the Q-fermion. There is no space-time within the BQUeST space. That situation is an advantage since we can choose the 7+1 dimension MOST (or UTMOST as shown in chapter 6) space-time as the space wherein the Q-fermion(s) reside.

We can further suggest that our universe emerged from an extremely high energy Q-fermion "seed" at the Big Bang point as described in section 2.3.1.

The connection between BQUeST and MOST (UTMOST) substantiates the role of BQUeST as the basis of QUeST and UST, and enhances the view that our universe lies within the Megaverse.

Most importantly, since BQUeST generates the 256 dimensions of a 32 complex quaternion space, BQUeST implies hypercomplex dimensions and coordinates.

5. A Deeper Origin for 64 Complex Octonion Dimension UTMOST Space

We found it necessary to expand MOST into a larger theory that we called UTMOST in Blaha (2020g) to obtain a more satisfactory basis for a Megaverse theory. The conceptual difficulty with MOST is that a deeper theory leading to it would require *two* fundamental fermions. We view this requirement as less than satisfactory for a truly fundamental basis.[18]

UTMOST is a generalization of MOST (with its 32 complex octonion dimensions) to a 64 complex octonion dimension theory. MOST has 512 dimensions. UTMOST has 1024 dimensions. Later we will see that UTMOST has a basis with one dimension and one fermion particle. The fermion particle exists in a 10 dimension space.[19]

Our goal in developing UTMOST, and its basis, which we will call BMOST,[20] was to create the simplest possible core theory from which the theory of the Megaverse is generated. The author feels this theory, BMOST, may be connected to a sibling theory such as a Superstring theory.

5.1 UTMOST Features

UTMOST has *64* complex octonion dimensions It is an extension of MOST that effectively doubles the number of dimensions, fermions and symmetries.

The "derivation" of the deeper origin of UTMOST again begins with fermion – dimension duality, in which the number of fermions equals the number of UTMOST dimensions: 1024. This equivalence again extends down to the individual particle and dimension level, and allows us to map between (fundamental) fermions and dimensions. We will call the basis of UTMOST, BMOST.

---

[18] See Blaha (2020g) for a description of the basis of MOST , which requires two fundamental fields.

[19] Interestingly some SuperString theories are 10 dimension. A connection of UTMOST to Superstrings may exist.

[20] Blaha (2020g) names these theories in a slightly different way.

5.2 UTMOST

UTMOST has a 64 dimension complex octonion space (Fig. 5.1) .

```
• • • • • • • •   • • • • • • • •
• • • • • • • •   • • • • • • • •
• • • • • • • •   • • • • • • • •
• • • • • • • •   • • • • • • • •
• • • • • • • •   • • • • • • • •
         ...
• • • • • • • •   • • • • • • • •
```

Figure 5.1a. The 64 dimension of UTMOST with 1024 dimensions.

We can reorder the UTMOST dimension array into a 32×32 array. We will find this format important in chapter 6.

```
• • • • • • • •   • • • • • • • •   • • • • • • • •   • • • • • • • •
• • • • • • • •   • • • • • • • •   • • • • • • • •   • • • • • • • •
• • • • • • • •   • • • • • • • •   • • • • • • • •   • • • • • • • •
• • • • • • • •   • • • • • • • •   • • • • • • • •   • • • • • • • •
                           ...
• • • • • • • •   • • • • • • • •   • • • • • • • •   • • • • • • • •
```

Figure 5.1b. The reordered UTMOST array with 32 dimensions and *double* complex octonion rows. It has a 32×32 form with 1024 dimensions.

We can divide the 32×32 dimension array into $4 \times 4 = 16$ dimension blocks as shown in Fig. 5.1c in a manner similar to Fig. 3.1. Each block has 16 dimensions. Each set of four blocks has 64 dimensions and has internal symmetries of the form of Fig. 2.7. We have separated the dimension array into eight sets: Normal+Dark1, Dark2+Dark3, Dark4+Dark5, and Dark6+Dark7. See Fig. 5.1c.

Normal + Dark1		Dark2 + Dark3		Dark4 + Dark5		Dark6 + Dark7	
4	4	4	4	4	4	4	4

Figure 5.1c. *Four* layers in 32 × 32 dimension array of 4 × 4 blocks. There are four-block 8 × 8 sections for each pair: Normal+Dark1, Dark2+Dark3, Dark4+Dark5 and Dark6+Dark7. In total they form the 32 × 32 = 1024 UTMOST dimension array. See Fig. 1.2 for the makeup of each 8 × 8 section.

5.3 UTMOST Symmetries

32 × 32 UTMOST can be viewed as 4 layers (copies), each with the internal symmetries of Fig. 5.2. The 4 layers are represented by 8 rows in Fig. 5.1c.

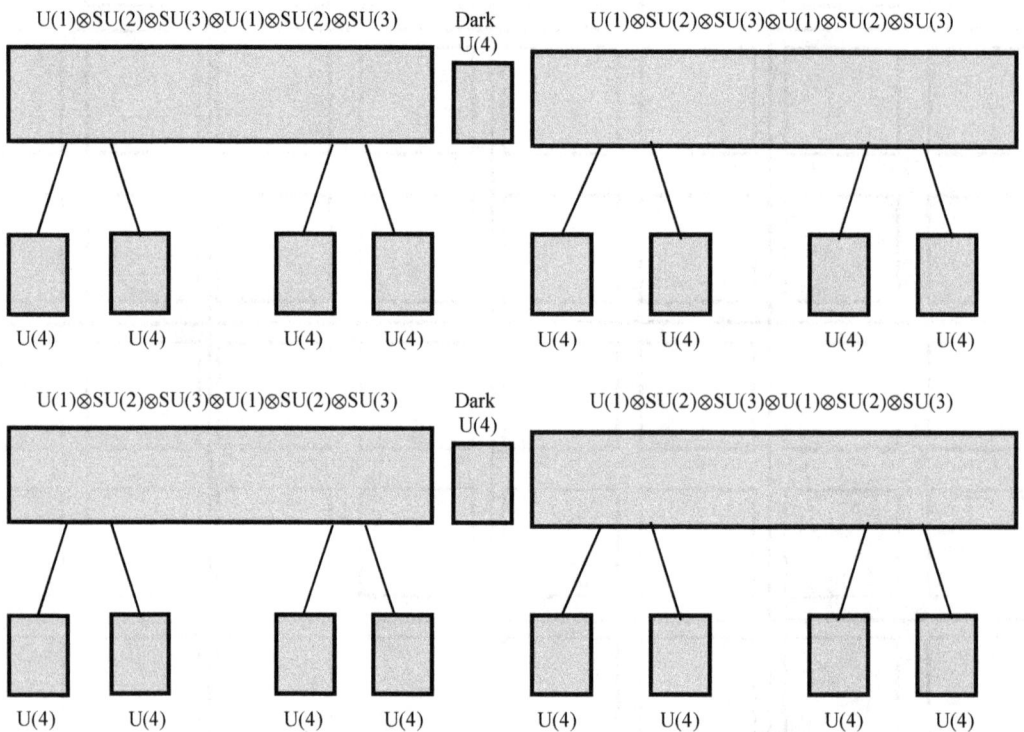

Figure 5.2. The internal symmetry groups of *one* of the four layers (8 rows of Fig. 5.1c) within the UTMOST 32 × 32 set of dimensions. The two "large" blocks

above are each a set of 20 dimensions furnishing fundamental representations of the indicated groups. The lower U(4) groups are Generation and Layer groups. The Dark U(4) group is shown. The total number of intenal symmetry dimensions in each layer copy is 224. The total number of internal symmetry dimensions in all four copies is 896. The difference 1024 − 896 = 128 dimensions gives eight complex octonion space-time dimensions. This figure corresponds to two rows of Fig 5.1c.

The 896 dimensions for the symmetry groups' fundamental representations:

$$[SU(2)\otimes U(1)\otimes SU(3)]^{32}\otimes U(4)^{72} \tag{5.1}$$

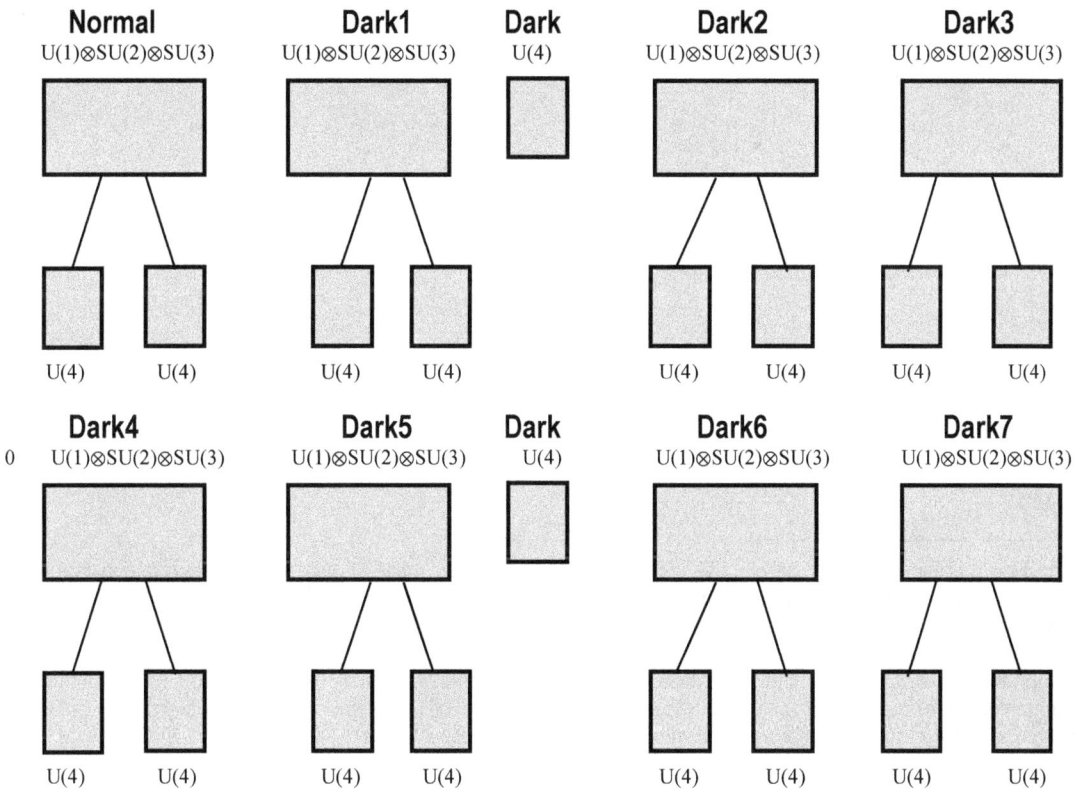

Figure 5.3. The internal symmetry groups of Fig. 5.2 expanded. This figure shows the internal symmetry groups of *one layer* (consisting of 8 complex octonion dimensions) of the four layers of 32 × 32 dimension UTMOST.

    The four layers of UTMOST are four copies of Fig. 5.3.[21] The factorization of the set of dimensions is accomplished by following the procedure given earlier.[22]

---

[21] The Layer groups are U(4) groups. They mix the generations of each of the top four layers, generation by generation, separately from the Layer groups mixing the lower four layers. This feature enables QUeST universes to be generated from either the top four layers or the lower four layers.

[22] The separation of the dimensions into the subgroup factors' representation can be implemented as group transformations and definitions using standard group theoretic methods. A more formal method for extracting the subgroup content of representations uses a symmetric group analysis of U(n) representation characters. See S. Blaha, J. Math. Phys. **10**, 2156 (1969) for a detailed discussion of this approach.

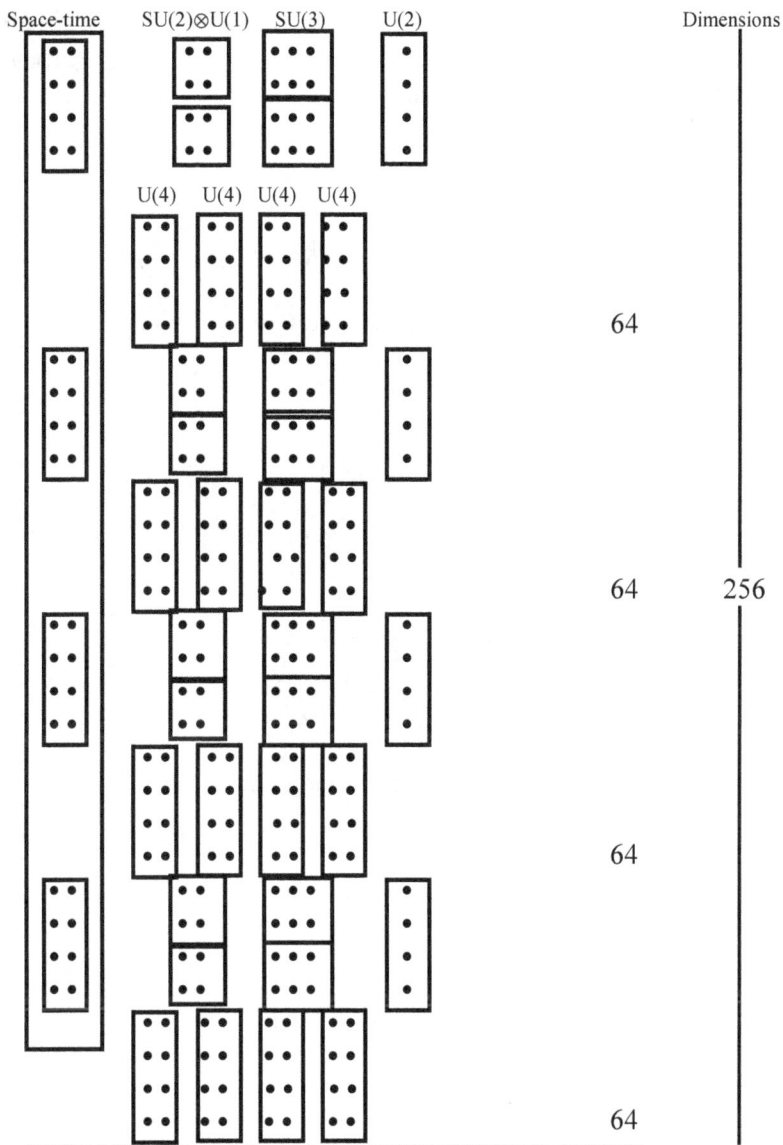

Figure 5.4 Two of the eight rows in Fig. 5.1c giving one of the four layers of 32 × 32 dimension UTMOST. The left long block is for 32 dimension space-time, which combines with the other three layers to give a 128 dimension space-time (an eight complex octonion dimension space-time.)

5.4 UTMOST Internal Symmetry Groups

The internal symmetry groups of UTMOST are four copies of the groups

"Normal" Gauge Groups
SU(3)⊗SU(2)⊗U(1)
Generation Group U(4)
Layer Group U(4)

Dark1 Gauge Groups
SU(3)⊗SU(2)⊗U(1)
Generation Group U(4)
Layer Group U(4)

Dark2 Gauge Groups
SU(3)⊗SU(2)⊗U(1)
Generation Group U(4)
Layer Group U(4)

Dark3 Gauge Groups
SU(3)⊗SU(2)⊗U(1)
Generation Group U(4)
Layer Group U(4)

Dark U(4) Group

Dark4 Gauge Groups
SU(3)⊗SU(2)⊗U(1)
Generation Group U(4)
Layer Group U(4)

Dark5 Gauge Groups
SU(3)⊗SU(2)⊗U(1)
Generation Group U(4)
Layer Group U(4)

Dark6 Gauge Groups
SU(3)⊗SU(2)⊗U(1)
Generation Group U(4)
Layer Group U(4)

Dark7 Gauge Groups
SU(3)⊗SU(2)⊗U(1)
Generation Group U(4)
Layer Group U(4)

Dark U(4) Group

Figure 5.5. One layer of UTMOST internal symmetry groups.

In addition to the above internal symmetry groups UTMOST allocates 128 dimensions to an 8 complex octonion space-time with 7 + 1 complex octonion coordinates. See Fig. 5.4.

5.5 UTMOST Fermions

The fermion and vector boson spectrums that emerge in UTMOST are those of an "enlarged" QUeST and UST (Unified SuperStandard Theory). The fermions are displayed below in Fig. 5.6. UTMOST has an additional six Dark sectors beyond QUeST and the Unified SuperStandard Theory (UST).

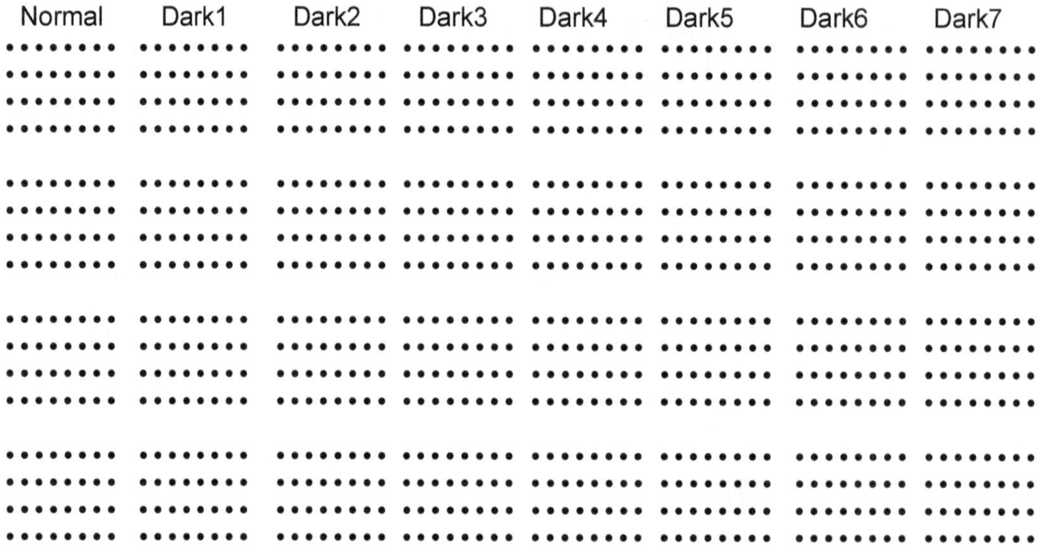

Normal	Dark1	Dark2	Dark3	Dark4	Dark5	Dark6	Dark7

Figure 5.6. Spectrum of UTMOST fermions in a 16×64 format. Each fermion is represented by a •. Including each quark. Each set of eight •.'s represents a charged lepton, a neutral lepton, three up-type quarks, and three down-type quarks. There are eight sets of four species in four generations which are in turn in 4 layers. There are 1024 fundamental fermions taking account of quark triplets. Note: Quark singlets won't do; triplets are required.

There are 1024 fermions and 1024 dimensions giving a match supporting fermion-dimension duality.

6. BMOST Derivation

This chapter derives a deeper basis for UTMOST[23] that we call BMOST.

6.1 Reordering of UTMOST Dimension Array

The UTMOST dimension array follows from the UTMOST 64 dimension complex octonion space (See Fig. 5.1). It is convenient to reorder[24] the array to the 32 × 32 form shown in Fig. 6.1.

Figure 6.1. The reordered UTMOST array having a 32 × 32 form with 1024 dimensions.

6.2 Derivation of UTMOST from BMOST with a One Dimension Space

We begin by assuming a one dimension space for BMOST, which we will call the *MQ-basis*. where the dimension has an associated dimension functional with arguments of spin and momentum but not internal symmetries explicitly.

To obtain a 32 × 32 array of dimensions we assume the MQ-basis dimension functional resides in a separate *ten* dimension space since the spin of the functional must have 32 components.

---

[23] Blaha (2020g) has a derivation of a deeper layer of MOST that we found to be less desirable.

[24] We do not use quaternion or octonion algebra in this book. Quaternions and octonions are only treated as the source of dimensions. It is possible to expand QUeST, MOST, and so on, to include consequences of quaternion and octonion algebra with, perhaps, a wider range of results.

We assume a single fermion, which we call an *MQ-fermion*, with two fermion fields,[25] each with 32-spinors. Treating the spinor components as independent[26] we can form a 32×32 array of functionals

$$D_{Mij} = f_M(p_1, s_{1i})f_M((p_2, s_{2j}) \tag{6.1}$$

which we take to be the functionals of UTMOST. The UTMOST dimension array is then

$$D_{MQij} = D_{Mij} \tag{6.2}$$

with composite functionals

$$f_{MQ}(p_1, p_2) \tag{6.3}$$

Thus we obtain the 32×32 UTMOST dimension array.

6.3 Ten Dimension Space and the Megaverse

The price of generating the UTMOST dimensions from a one dimension space is the introduction of a ten dimension space for a MQ-fermion with two PseudoQuantum fields. At first glance this situation is problematic. We have a one dimension MQ-basis with a corresponding MQ-fermion. However the ten dimension space is extraneous to the MQ-basis space. The UTMOST space-time is a 7+1-dimension complex octonion space-time.

The ten dimension space is extraneous to UTMOST and the MQ-basis. Given its tenfold nature and the common appearance of 10 dimension spaces in SuperString

---

[25] Blaha's PseudoQuantum formulation of Quantum Field theory has two fields representing each fermion and boson particle. It has the advantages of allowing a canonical formulation of quantum field theory with higher order differential equations such as those needed for quark confieement, and of allowing quantum field theory formulations in *any* space-time. See S. Blaha, Phys Rev **D10**, 4268 (1974); _____, **D11**, 2921 (1974) for PseudoQuantum quark confinement with an r potential. See S. Blaha, Phys. Rev **D17**, 994 (1978), ____, IL Nuovo Cimento **49A**, 35 (1979), _____, IL Nuovo Cimento **49A**, 113 (1979) for PseudoQuantum formulation in arbitrary space-times using Bogoliubov transformations..

[26] The spinor components of each fermion field is independent of the others field's spinor components.

theories one is tempted to suggest a connection. One could view a SuperString theory as the root of the Megaverse producing a Megaverse seed that becomes the Megaverse, which in turn generates universes. The issue remains to be studied.

6.4 Iterative Generation of UTMOST Dimensions

We can also establish the 1024 dimension UTMOST array in a two-step process. First we define an MQ-fermion with a functional that has 4-spinor components in a 3+1 exterior space-time. Each component of this off-shell Dirac field functional is independent. We then define the two MQ-fermion functionals

$$f_{M1}(p, s) \quad \text{and} \quad f_{M2}(p, s) \qquad (6.4)$$

with each functional, $f_{M1}(p, s)$ and $f_{M2}(p, s)$, having four spinor components. Thus their product forms a 4×4 array.

$$F_M(i, j) = f_{M1}(p, s_i) \, f_{M2}(p, s_j) \qquad (6.5)$$

The indices i, j = 1, 2, 3, 4 will subsequently label 4×4 blocks of functionals.

Next, this 4×4 dimension (block) array generates the UTMOST 32×32 array of dimensions by viewing each generated functional $F_M(i, j)$ as describing a block of $4 \times 4 = 16$ dimensions. This is done by treating each $F_M(i, j)$ block as the product of two Dirac functionals of the same form as eq. 6.4 with 8-spinor components. Fig. 6.3 shows the process of generating UTMOST from the MW-basis.

Mapping to dimensions gives the UTMOST array of $(4 \times 8)^2 = 1024$ dimensions. This map to a 32×32 dimension array is diagrammed in Fig. 6.2.

6.4.1 New View of Block Structure of UTMOST array

The intermediate form of 4×4 dimension arrays suggests a new form of structuring the 1024 dimension UTMOST array as composed of 64 blocks of $4 \times 4 = 16$ dimensions. The blocks would each correspond to a U(8) symmetry.

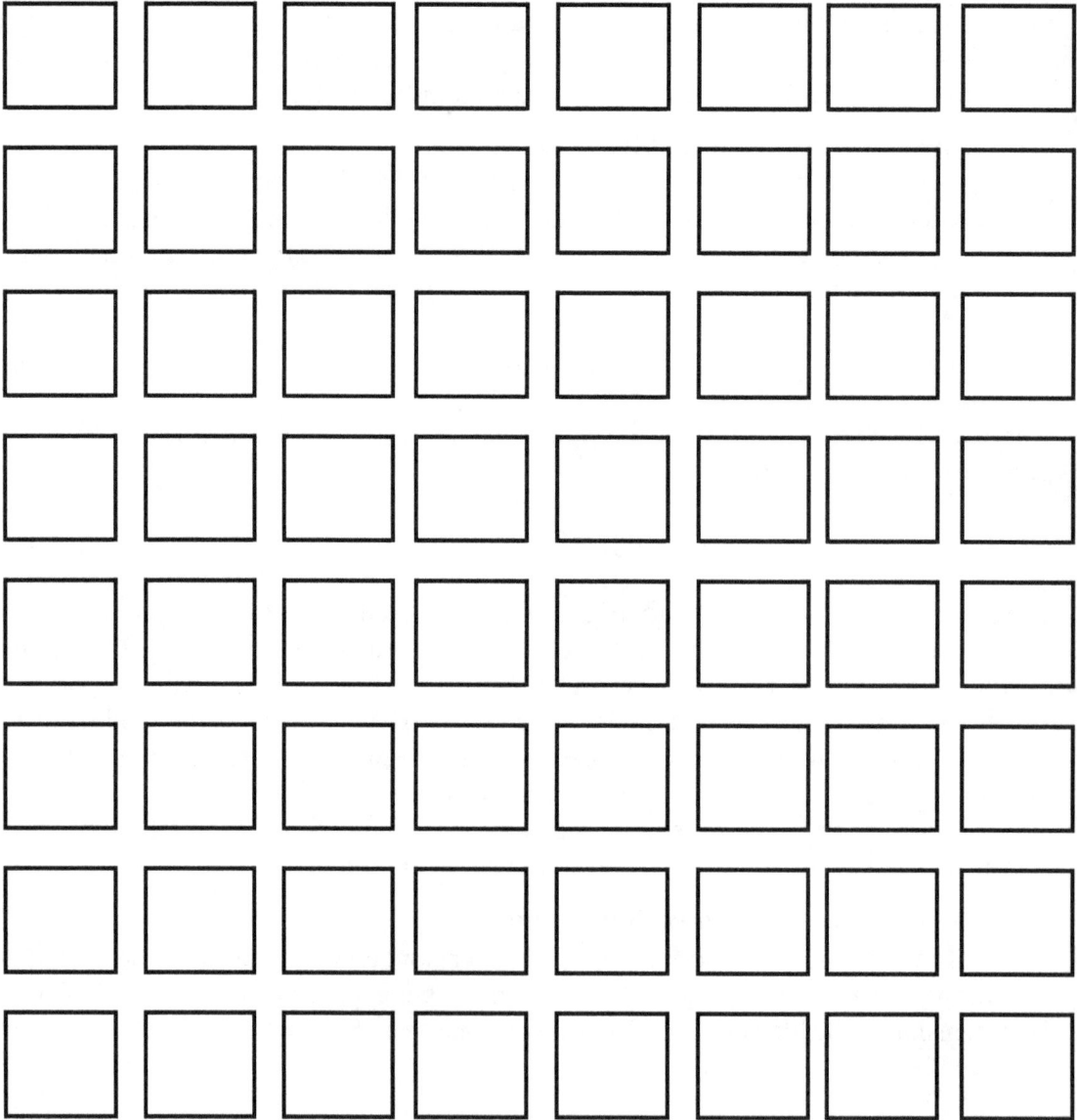

Figure 6.2. Block form of 32 × 32 = 1024 UTMOST dimension array with each pair of rows corresponding to one layer of 4 layers.

The block form of Fig. 6.2 has significant implications for the internal symmetry structure of the UTMOST array and for the structure of the 1024 set of fermions.

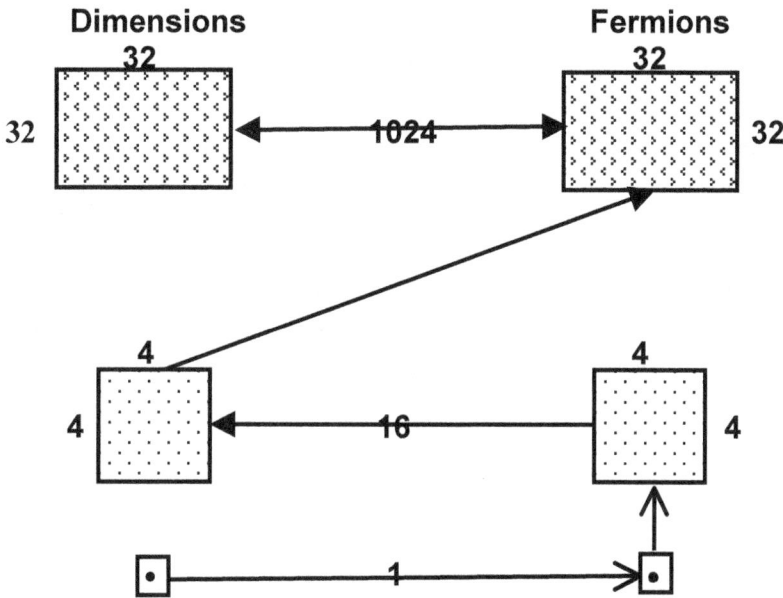

Figure 6.3. Diagram of the ascent from the MQ-basis to the UTMOST array.

6.5 Known Symmetry Group Structure

Fig. 5.3 displays the form of one layer of the UTMOST dimension array with subgroups indicated by boxed dimensions. (There are eight layers in UTMOST 16 × 64 format.)

Fig. 5.1c shows the complete set of eight layers of UTMOST dimensions numbering 1024 dimensions in total.

6.6 Fermion Structure

The overall structure of UTMOST fermions is displayed in Fig. 5.6. The fermions can be reordered to display a 4×4 block structure as suggested by Fig. 6.2.

Fig. 6.4 displays a set of sixteen 4×4 blocks with each block holding 16 fermions. The fermions are distributed among the blocks in a pattern similar to Fig. 5.6 with Normal fermions in the first two columns and Dark fermions in the remaining columns. Note: the fermions are reordered. The 1st, 3rd, 5th, and 7th blocks have a charged lepton paired with an up-type quark. The other blocks have a neutral lepton paired with a down-type quark.

6.7 Fermion Species and the Pairing of Fermions in Blocks

In our development of UST in Blaha (2020c) and earlier books we introduced the idea of four types of fermion species: charged lepton, neutral lepton, up-type quark, and down-type quark. We suggested that fermion species were a result of the application of four types of Complex Lorentz boosts. We found that it was natural to identify charged leptons as Dirac fermions, neutral leptons as tachyon fermions,[27] up-type quarks as Dirac-type fermions with complex 3-momentum, and down-type quarks[28] as tachyon fermions with complex 3-momentum.

The pairing of the leptons and quarks, which follows from the need to have 4×4 blocks, is based on pairing Dirac-like fermions in one type of block, and tachyon fermions in the other type of block in Fig. 6.4.

---

[27] Blaha (2020c) describes experimental data consistent with tachyon neutrinos.
[28] This possibility may explain 'x' cutoffs, and other modifications, needed to make parton calculations agree with deep inelastic scattering data.

Normal				Dark1				Dark2				Dark3			
e q-up		v q-down		e q-up		v q-down		e q-up		v q-down		e q-up		v q-down	
4		4		4		4		4		4		4		4	

Dark4		Dark5		Dark6		Dark7	

Figure 6.4. Block form of the 32 × 32 UTMOST fermion array with each row corresponding to *half of a layer*. (Compare to Fig. 1.12.) Thus 8 × ½ = 4 layers results. Each block contains four generations of fermions. The result is sixty-four 4 × 4 blocks. The label e q-up indicates a charged lepton – up-type quark pair, v q-down indicates a neutral lepton – down-type quark pair, and so on. *The form displayed here explains why generations come in fours.*

6.8 Reordered Fermion Structure Implications

A number of implications follow from the form of Fig. 6.4:

1. Each block has a quartet that is similar in form to a 3+1 coordinate system. This similarity again raises the possibility of a Lorentz-like transformation group where a lepton plays the role of time and a quark triplet plays the role of spatial coordinates. A form of this possibility was considered in Blaha (2020c). It also raises the possibility of symmetry breaking along the lines of a Lorentz group.

2. The SU(2)⊗U(1) ElectroWeak group structure is clearly evident in the fermion blocks. Fig. 6.5 shows the charged generators of the group act between the first and second column blocks, and between the third and fourth column blocks (the horizontal arrows). The neutral charge group generators act vertically in each column (vertical arrows). ElectroWeak transformations can change ElectroWeak quantum numbers. They are diagonal in other particle internal quantum numbers.

Normal

e q-up   v q-down

Dark

e q-up   v q-down

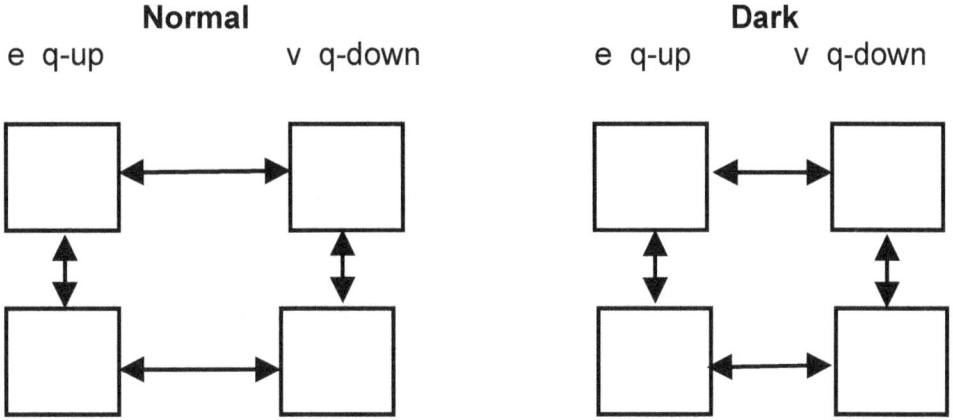

Figure 6.5. The effect of the ElectroWeak generator fields on the blocks of fermions. Charged generators of the group act between the first and second column blocks, and between the third and fourth column blocks (the horizontal arrows). The neutral charge group generators acts vertically in each column (vertical arrows).

6.9 UTMOST Dimension Restructure

We now will restructure the dimension array of Fig. 6.1 from the perspective of the 4 ×4 block structure suggested by the two-step derivation (section 6.4) of UMOST dimensions from BMOST.

6.9.1 An Attempt at UTMOST 16 dimension Blocks

If we develop a block structure for UTMOST along the lines of our QUeST block structure development, we encounter a problem with the U(4) Dark group. It does not allow us to put UTMOST into a block form similar to that of QUeST in Fig. 2.7. Fig. 2.7 nicely embeds a U(2) Dark group in an SU(3) block. Placing a U(4) in a similar SU(3) block for UTMOST yields an unsatisfactory 20 dimension block.

6.10 Factorized UTMOST: Sixteen Dimension Blocks

The only apparent "simple" way to have a 16 dimension blocks for the UTMOST array is to factor the set of dimensions into two parts: one part containing the Normal and Dark1 set of dimensions, and the other block containing the Dark2 and Dark3 set of dimensions, and so on. See Fig. 6.6.

Then we cant also factor UTMOST space into two 512 dimension parts. (See section 5.3 for the UTMOST space-time discussion.) *The result is a separated UTMOST space-time with each having the 64 dimensions of an eight complex quaternion space-time.*

Fig. 6.6 shows one layer (out of four layers) of the UTMOST dimension array separated using 16 dimension blocks. Thus the intermediate stage of the derivation (section 6.4.1) is realized in 16 dimension blocks.

The four layers of UTMOST are copies of Fig. 6.6. Each of the layers has its own set of internal symmetry interactions. The 64 space-time dimensions in each factor lead to a 7+1 complex quaternion space-time.

Each of the two parts of UTMOST is a MOST representation.

6.10.1 Two Space-Times Factors

The two space-times obtained by separation can be viewed as orthogonal. Each allows separate Lorentz-like transformations. A combined Lorentz-like transformation of both space-times is not supported. (One could add a combined Lorentz-like set of transformations if it were justified.)

The two space-times, having separate metric tensors, would appear to have separate theories of General Relativity. The introduction of a global space-time containing both space-time factors would lead to a universal Megaverse Theory of Gravitation.

6.10.2 Factorized UTMOST Fermions

Factorizing UTMOST divides the fermion spectrum into two 512 fermion parts.

A. Normal and Dark1 Parts

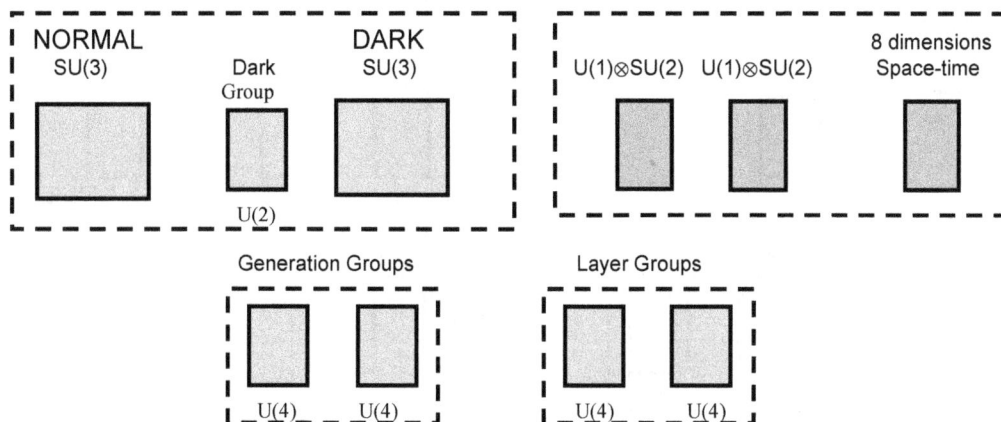

B. Dark2 and Dark3 Parts

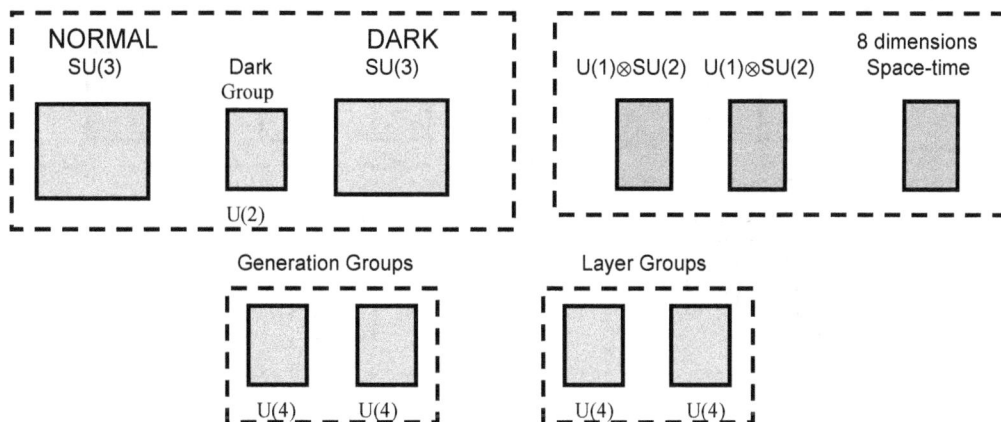

C. Dark4 and Dark5 Parts

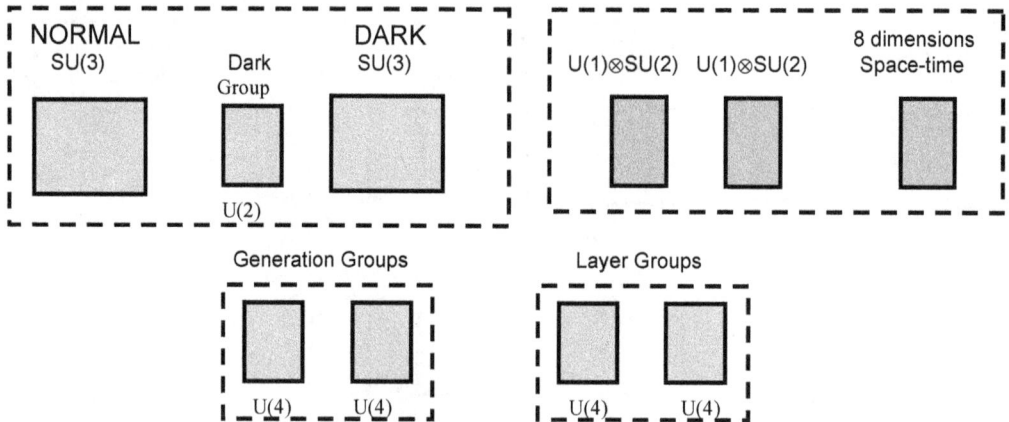

D. Dark6 and Dark7 Parts

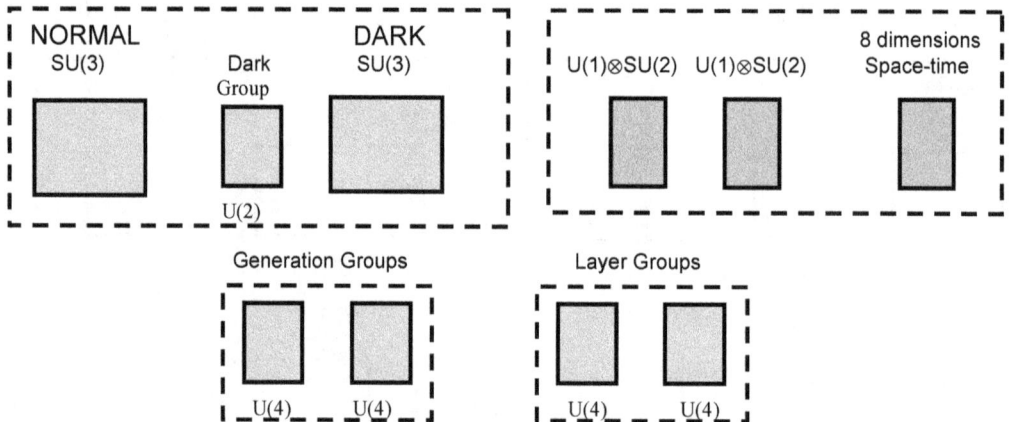

Figure 6.6. Separating internal symmetry groups for *one* UTMOST layer. Including space-time. Each "dotted" block (regardless of size) contains 16 dimensions. The U(2) groups supports transformations (rotations) between Normal and Dark1, *and* Dark2 and Dark3, and so on *separately*.

	Normal			**Dark1**			**Dark2**			**Dark3**	
	e q-up	v q-down		e q-up	v q-down		e q-up	v q-down		e q-up	v q-down
	4	4		4	4		4	4		4	4
4											
4											
4											
4											

	Dark4			**Dark5**			**Dark6**			**Dark7**	
4											
4											
4											
4											

Figure 6.7. Block form of 32 × 32 UTMOST fermion array showing separated fermion arrays with 512 fermions based on Fig. 6.4. Each pair of rows corresponds to one layer. Each block contains four generations of fermions. The result is 4 × 4 blocks. The label e q-up indicates a charged lepton – up-type quark pair, ν q-down indicates a neutral lepton – down-type quark pair, and so on.

6.11 Some Implications of the New UTMOST Dimension Array Structure

The 1024 UTMOST dimensions imply an initial fundamental representation of U(512). The block structure of UTMOST (with Fig. 6.6 expanded to eight layers) suggest that there are two breakdowns: one breakdown to $U(8)^{64}$, and a "fine structure-like" breakdown to the internal symmetry groups of Fig. 6.6.

The subdivision into blocks of 16 dimensions has a number of implications:

1. The separation into two sets of symmetries implies each part of UTMOST has MOST dimensions, fermions, and internal symmetries.

2. It suggests a set of subgroups that may be relevant for symmetry breaking. The combination of U(1)⊗SU(2) groups with space-time (before further breakdown) was presaged by a study of the relation of ElectroWeak symmetry to complex space-time in Blaha (2020c) and in earlier books by the author.

3. The combination of the SU(3) groups and the U(2) Dark group seems quite natural at the time before "fine structure" symmetry breakdown. (The Dark group rotates Normal and Dark fermions with the same U(1)⊗SU(2)⊗SU(3) internal symmetry indices.)

4. The block with SU(3) and U(2) groups represents the combined internal symmetries before symmetry breakdown.

5. The Generation groups are combined as they should be before symmetry breaking. Similarly the Layer groups are combined.

We conclude that the block pattern of Fig. 6.6 and its eight layer UTMOST pattern are in accord with two symmetry breakdowns: a breakdown to $U(8)^{64}$, and a "fine structure-like" breakdown to the groups depicted in Fig. 6.6.

We have derived the complex octonion dimensions of UTMOST from BMOST, and found a remarkable breakdown of U(512) to blocks with U(8) symmetries.

7. Summary of QUeST, BQUeST, MOST, UTMOST, and BMOST Features

This chapter summarizes the features of the above theories: their space, their dimensions, their fermions, their space-times, and any associated spaces.

	Dimensions	Fermions	Space-Time
32 Complex Quaternion QUeST	256	256	3+1 Complex Quaternion
1 Dimension BQUeST	1	1	None
Associated Space-Time: 7+1 Complex Quaternion MOST			
32 Complex Octonion MOST	512	512	7+1 Complex Quaternion
64 Complex Octonion UTMOST	1024	1024	7 + 1 Complex Octonion
1 Dimension BMOST	1	1	None
Associated Space-Time: External 10 dimensions (possibly Complex Octonion)			
Each UTMOST Factor	512	512	7+1 Complex Quaternion

Figure 7.1. Comparison of QUeST, BQUeST, MOST, UTMOST, and BMOST.

7.1 Some Implications

1. BQUeST space-time can be viewed as residing in MOST. Thus we can have QUeST begin as a BQUeST seed within MOST.

2. UTMOST can be viewed as composed of two separate parts that are each MOST spaces.

REFERENCES

Akhiezer, N. I., Frink, A. H. (tr), 1962, *The Calculus of Variations* (Blaisdell Publishing, New York, 1962).

Bjorken, J. D., Drell, S. D., 1964, *Relativistic Quantum Mechanics* (McGraw-Hill, New York, 1965).

Bjorken, J. D., Drell, S. D., 1965, *Relativistic Quantum Fields* (McGraw-Hill, New York, 1965).

Blaha, S., 1998, *Cosmos and Consciousness* (Pingree-Hill Publishing, Auburn, NH, 1998).

_____, 2002, *A Finite Unified Quantum Field Theory of the Elementary Particle Standard Model and Quantum Gravity Based on New Quantum Dimensions™ & a New Paradigm in the Calculus of Variations* (Pingree-Hill Publishing, Auburn, NH, 2002).

_____, 2003, *A Finite Unified Quantum Field Theory of the Elementary Particle Standard Model and Quantum Gravity Based on New Quantum Dimensions™ and a New Paradigm in the Calculus of Variations* (Pingree-Hill Publishing, Auburn, NH, 2003).

_____, 2004, *Quantum Big Bang Cosmology: Complex Space-time General Relativity, Quantum Coordinates™Dodecahedral Universe, Inflation, and New Spin 0, ½, 1 & 2 Tachyons & Imagyons* (Pingree-Hill Publishing, Auburn, NH, 2004).

_____, 2005a, *Quantum Theory of the Third Kind: A New Type of Divergence-free Quantum Field Theory Supporting a Unified Standard Model of Elementary Particles and Quantum Gravity based on a New Method in the Calculus of Variations* (Pingree-Hill Publishing, Auburn, NH, 2005).

_____, 2005b, *The Metatheory of Physics Theories, and the Theory of Everything as a Quantum Computer Language* (Pingree-Hill Publishing, Auburn, NH, 2005).

112   *REFERENCES*

_____, 2005c, *The Equivalence of Elementary Particle Theories and Computer Languages: Quantum Computers, Turing Machines, Standard Model, Superstring Theory, and a Proof that Gödel's Theorem Implies Nature Must Be Quantum* (Pingree-Hill Publishing, Auburn, NH, 2005).

_____, 2006a, *The Foundation of the Forces of Nature* (Pingree-Hill Publishing, Auburn, NH, 2006).

_____, 2006b, *A Derivation of ElectroWeak Theory based on an Extension of Special Relativity; Black Hole Tachyons; & Tachyons of Any Spin.* (Pingree-Hill Publishing, Auburn, NH, 2006).

_____, 2007a, *Physics Beyond the Light Barrier: The Source of Parity Violation, Tachyons, and A Derivation of Standard Model Features* (Pingree-Hill Publishing, Auburn, NH, 2007).

_____, 2007b, *The Origin of the Standard Model: The Genesis of Four Quark and Lepton Species, Parity Violation, the ElectroWeak Sector, Color SU(3), Three Visible Generations of Fermions, and One Generation of Dark Matter with Dark Energy* (Pingree-Hill Publishing, Auburn, NH, 2007).

_____, 2008a, *A Direct Derivation of the Form of the Standard Model From GL(16) (Pingree-Hill Publishing, Auburn, NH, 2008).*

_____, 2008b, *A Complete Derivation of the Form of the Standard Model With a New Method to Generate Particle Masses Second Edition* (Pingree-Hill Publishing, Auburn, NH, 2008)

_____, 2009, *The Algebra of Thought & Reality: The Mathematical Basis for Plato's Theory of Ideas, and Reality Extended to Include A Priori Observers and Space-Time Second Edition* (Pingree-Hill Publishing, Auburn, NH, 2009).

_____, 2010a, *Operator Metaphysics: A New Metaphysics Based on a New Operator Logic and a New Quantum Operator Logic that Lead to a Mathematical Basis for Plato's Theory of Ideas and Reality* (Pingree-Hill Publishing, Auburn, NH, 2010).

_____, 2010b, *The Standard Model's Form Derived from Operator Logic, Superluminal Transformations and GL(16)* (Pingree-Hill Publishing, Auburn, NH, 2010).

_____, 2010c, *SuperCivilizations: Civilizations as Superorganisms* (McMann-Fisher Publishing, Auburn, NH, 2010).

_____, 2011a, *21st Century Natural Philosophy Of Ultimate Physical Reality* (McMann-Fisher Publishing, Auburn, NH, 2011).

_____, 2011b, *All the Universe! Faster Than Light Tachyon Quark Starships & Particle Accelerators with the LHC as a Prototype Starship Drive Scientific Edition* (Pingree-Hill Publishing, Auburn, NH, 2011).

_____, 2011c, *From Asynchronous Logic to The Standard Model to Superflight to the Stars* (Blaha Research, Auburn, NH, 2011).

_____, 2012a, *From Asynchronous Logic to The Standard Model to Superflight to the Stars volume 2: Superluminal CP and CPT, U(4) Complex General Relativity and The Standard Model, Complex Vierbein General Relativity, Kinetic Theory, Thermodynamics* (Blaha Research, Auburn, NH, 2012).

_____, 2012b, *Standard Model Symmetries, And Four And Sixteen Dimension Complex Relativity; The Origin Of Higgs Mass Terms* (Blaha Reasearch, Auburn, NH, 2012).

_____, 2013a, *Multi-Stage Space Guns, Micro-Pulse Nuclear Rockets, and Faster-Than-Light Quark-Gluon Ion Drive Starships* (Blaha Research, Auburn, NH, 2013).

_____, 2013b, *The Bridge to Dark Matter; A New Sister Universe; Dark Energy; Inflatons; Quantum Big Bang; Superluminal Physics; An Extended Standard Model Based on Geometry* (Blaha Reasearch, Auburn, NH, 2013).

_____, 2014a, *Universes and Megaverses: From a New Standard Model to a Physical Megaverse; The Big Bang; Our Sister Universe's Wormhole; Origin of the Cosmological Constant, Spatial Asymmetry of the Universe, and its Web of Galaxies; A Baryonic Field between Universes and Particles; Megaverse Extended Wheeler-DeWitt Equation* (Blaha Reasearch, Auburn, NH, 2014).

_____, 2014b, *All the Megaverse! Starships Exploring the Endless Universes of the Cosmos Using the Baryonic Force* (Blaha Research, Auburn, NH, 2014).

_____, 2014c, *All the Megaverse! II Between Megaverse Universes: Quantum Entanglement Explained by the Megaverse Coherent Baryonic Radiation Devices – PHASERs Neutron Star Megaverse Slingshot Dynamics Spiritual and UFO Events, and the Megaverse Microscopic Entry into the Megaverse* (Blaha Research, Auburn, NH, 2014).

_____, 2015a, *PHYSICS IS LOGIC PAINTED ON THE VOID: Origin of Bare Masses and The Standard Model in Logic, U(4) Origin of the Generations, Normal and Dark Baryonic Forces, Dark Matter, Dark Energy, The Big Bang, Complex General Relativity, A Megaverse of Universe Particles* (Blaha Research, Auburn, NH, 2015).

_____, 2015b, *PHYSICS IS LOGIC Part II: The Theory of Everything, The Megaverse Theory of Everything, U(4)$\otimes$U(4) Grand Unified Theory (GUT), Inertial Mass = Gravitational Mass, Unified Extended Standard Model and a New Complex General Relativity with Higgs Particles, Generation Group Higgs Particles* (Blaha Research, Auburn, NH, 2015).

_____, 2015c, *The Origin of Higgs ("God") Particles and the Higgs Mechanism: Physics is Logic III, Beyond Higgs – A Revamped Theory With a Local Arrow of Time, The Theory of Everything Enhanced, Why Inertial Frames are Special, Universes of the Mind* (Blaha Research, Auburn, NH, 2015).

_____, 2015d, *The Origin of the Eight Coupling Constants of The Theory of Everything: U(8) Grand Unified Theory of Everything (GUTE), S^8 Coupling Constant Symmetry, Space-Time Dependent Coupling Constants, Big Bang Vacuum Coupling Constants, Physics is Logic IV* (Blaha Research, Auburn, NH, 2015).

_____, 2016a, *New Types of Dark Matter, Big Bang Equipartition, and A New U(4) Symmetry in the Theory of Everything: Equipartition Principle for Fermions, Matter is 83.33% Dark,*

Penetrating the Veil of the Big Bang, Explicit QFT Quark Confinement and Charmonium, Physics is Logic V (Blaha Research, Auburn, NH, 2016).

_____, 2016b, *The Periodic Table of the 192 Quarks and Leptons in The Theory of Everything: The U(4) Layer Group, Physics is Logic VI* (Blaha Research, Auburn, NH, 2016).

_____, 2016c, *New Boson Quantum Field Theory, Dark Matter Dynamics, Dark Matter Fermion Layer Mixing, Genesis of Higgs Particles, New Layer Higgs Masses, Higgs Coupling Constants, Non-Abelian Higgs Gauge Fields, Physics is Logic VII* (Blaha Research, Auburn, NH, 2016).

_____, 2016d, *Unification of the Strong Interactions and Gravitation: Quark Confinement Linked to Modified Short-Distance Gravity; Physics is Logic VIII* (Blaha Research, Auburn, NH, 2016).

_____, 2016e, *MoND: Unification of the Strong Interactions and Gravitation II, Quark Confinement Linked to Large-Scale Gravity, Physics is Logic IX* (Blaha Research, Auburn, NH, 2016).

_____, 2016f, *CQ Mechanics: A Unification of Quantum & Classical Mechanics, Quantum/Semi-Classical Entanglement, Quantum/Classical Path Integrals, Quantum/Classical Chaos* (Blaha Research, Auburn, NH, 2016).

_____, 2016g, *GEMS: Unified Gravity, ElectroMagnetic and Strong Interactions: Manifest Quark Confinement, A Solution for the Proton Spin Puzzle, Modified Gravity on the Galactic Scale* (Pingree Hill Publishing, Auburn, NH, 2016).

_____, 2016h, *Unification of the Seven Boson Interactions based on the Riemann-Christoffel Curvature Tensor* (Pingree Hill Publishing, Auburn, NH, 2016).

_____, 2017a, *Unification of the Eleven Boson Interactions based on 'Rotations of Interactions'* (Pingree Hill Publishing, Auburn, NH, 2017).

_____, 2017b, *The Origin of Fermions and Bosons, and Their Unification* (Pingree Hill Publishing, Auburn, NH, 2017).

116   *REFERENCES*

_____, 2017c, *Megaverse: The Universe of Universes* (Pingree Hill Publishing, Auburn, NH, 2017).

_____, 2017d, *SuperSymmetry and the Unified SuperStandard Model* (Pingree Hill Publishing, Auburn, NH, 2017).

_____, 2017e, *From Qubits to the Unified SuperStandard Model with Embedded SuperStrings: A Derivation* (Pingree Hill Publishing, Auburn, NH, 2017).

_____, 2017f, *The Unified SuperStandard Model in Our Universe and the Megaverse: Quarks, ... ,* (Pingree Hill Publishing, Auburn, NH, 2017).

_____, 2018a, *The Unified SuperStandard Model and the Megaverse SECOND EDITION A Deeper Theory based on a New Particle Functional Space that Explicates Quantum Entanglement Spookiness (Volume 1)* (Pingree Hill Publishing, Auburn, NH, 2018).

_____, 2018b, *Cosmos Creation: The Unified SuperStandard Model, Volume 2, SECOND EDITION* (Pingree Hill Publishing, Auburn, NH, 2018).

_____, 2018c, *God Theory (*Pingree Hill Publishing, Auburn, NH, 2018).

_____, 2018d, *Immortal Eye: God Theory: Second Edition* (Pingree Hill Publishing, Auburn, NH, 2018).

_____, 2018e, *Unification of God Theory and Unified SuperStandard Model THIRD EDITION* (Pingree Hill Publishing, Auburn, NH, 2018).

_____, 2019a, *Calculation of: QED α = 1/137, and Other Coupling Constants of the Unified SuperStandard Theory* (Pingree Hill Publishing, Auburn, NH, 2019).

_____, 2019b, *Coupling Constants of the Unified SuperStandard Theory SECOND EDITION* (Pingree Hill Publishing, Auburn, NH, 2019).

_____, 2019c, *New Hybrid Quantum Big_Bang–Megaverse_Driven Universe with a Finite Big Bang and an Increasing Hubble Constant* (Pingree Hill Publishing, Auburn, NH, 2019).

_____, 2019d, *The Universe, The Electron and The Vacuum* (Pingree Hill Publishing, Auburn, NH, 2019).

_____, 2019e, *Quantum Big Bang – Quantum Vacuum Universes (Particles)* (Pingree Hill Publishing, Auburn, NH, 2019).

_____, 2019f, *The Exact QED Calculation of the Fine Structure Constant Implies ALL 4D Universes have the Same Physics/Life Prospects* (Pingree Hill Publishing, Auburn, NH, 2019).

_____, 2019g, *Unified SuperStandard Theory and the SuperUniverse Model: The Foundation of Science* (Pingree Hill Publishing, Auburn, NH, 2019).

_____, 2020a, *Quaternion Unified SuperStandard Theory (The QUeST) and Megaverse Octonion SuperStandard Theory (MOST)* (Pingree Hill Publishing, Auburn, NH, 2020).

_____, 2020b, *United Universes Quaternion Universe - Octonion Megaverse* (Pingree Hill Publishing, Auburn, NH, 2020).

_____, 2020c, *Unified SuperStandard Theories for Quaternion Universes & The Octonion Megaverse* (Pingree Hill Publishing, Auburn, NH, 2020).

_____, 2020d, *The Essence of Eternity: Quaternion & Octonion SuperStandard Theories* (Pingree Hill Publishing, Auburn, NH, 2020).

_____, 2020e, *The Essence of Eternity II* (Pingree Hill Publishing, Auburn, NH, 2020).

_____, 2020f, *A Very Conscious Universe* (Pingree Hill Publishing, Auburn, NH, 2020).

_____, 2020g, *Hypercomplex Universe* (Pingree Hill Publishing, Auburn, NH, 2020).

_____, 2020h, *Beneath the Quaternion Universe* (Pingree Hill Publishing, Auburn, NH, 2020).

Eddington, A. S., 1952, *The Mathematical Theory of Relativity* (Cambridge University Press, Cambridge, U.K., 1952).

Fant, Karl M., 2005, *Logically Determined Design: Clockless System Design With NULL Convention Logic* (John Wiley and Sons, Hoboken, NJ, 2005).

Feinberg, G. and Shapiro, R., 1980, *Life Beyond Earth: The Intelligent Earthlings Guide to Life in the Universe* (William Morrow and Company, New York, 1980).

Gelfand, I. M., Fomin, S. V., Silverman, R. A. (tr), 2000, *Calculus of Variations* (Dover Publications, Mineola, NY, 2000).

Giaquinta, M., Modica, G., Souchek, J., 1998, *Cartesian Coordinates in the Calculus of Variations* Volumes I and II (Springer-Verlag, New York, 1998).

Giaquinta, M., Hildebrandt, S., 1996, *Calculus of Variations* Volumes I and II (Springer-Verlag, New York, 1996).

Gradshteyn, I. S. and Ryzhik, I. M., 1965, *Table of Integrals, Series, and Products* (Academic Press, New York, 1965).

Heitler, W., 1954, *The Quantum Theory of Radiation* (Claendon Press, Oxford, UK, 1954).

Huang, Kerson, 1992, *Quarks, Leptons & Gauge Fields 2nd Edition* (World Scientific Publishing Company, Singapore, 1992).

Jost, J., Li-Jost, X., 1998, *Calculus of Variations* (Cambridge University Press, New York, 1998).

Kaku, Michio, 1993, *Quantum Field Theory*, (Oxford University Press, New York, 1993).

Kirk, G. S. and Raven, J. E., 1962, *The Presocratic Philosophers* (Cambridge University Press, New York, 1962).

Landau, L. D. and Lifshitz, E. M., 1987, *Fluid Mechanics 2nd Edition*, (Pergamon Press, Elmsford, NY, 1987).

Misner, C. W., Thorne, K. S., and Wheeler, J. A., 1973, *Gravitation* (W. H. Freeman, New York, 1973).

Rescher, N., 1967, *The Philosophy of Leibniz* (Prentice-Hall, Englewood Cliffs, NJ, 1967).

Rieffel, Eleanor and Polak, Wolfgang, 2014, *Quantum Computing* (MIT Press, Cambridge, MA, 2014).

Riesz, Frigyes and Sz.-Nagy, Béla, 1990, *Functional Analysis* (Dover Publications, New York, 1990).

Sagan, H., 1993, *Introduction to the Calculus of Variations* (Dover Publications, Mineola, NY, 1993).

Sakurai, J. J., 1964, *Invariance Principles and Elementary Particles* (Princeton University Press, Princeton, NJ, 1964).

Streater, R. F. and Wightman, A. S., 2000, *PCT, Spin, Statistics, and All That* (Princeton University Press, Princeton, NJ 2000).

Weinberg, S., 1972, *Gravitation and Cosmology* (John Wiley and Sons, New York, 1972).

Weinberg, S., 1995, *The Quantum Theory of Fields Volume I* (Cambridge University Press, New York, 1995).

Weinberg, S., 2000, *The Quantum Theory of Fields Volume III Supersymmetry* (Cambridge University Press, New York, 2000).

Weyl, H., 1950, *Space, Time, Matter* (Dover, New York, 1950).

Weyl, H., (Tr. S. Pollard et al), 1987, *The Continuum* (Dover Publications, New York, 1987).

REFERENCES

INDEX

About the Author

Stephen Blaha is a well-known Physicist and Man of Letters with interests in Science, Society and civilization, the Arts, and Technology. He had an Alfred P. Sloan Foundation scholarship in college. He received his Ph.D. in Physics from Rockefeller University. He has served on the faculties of several major universities. He was also a Member of the Technical Staff at Bell Laboratories, a manager at the Boston Globe Newspaper, a Director at Wang Laboratories, and President of Blaha Software Inc. and of Janus Associates Inc. (NH).

Among other achievements he was a co-discoverer of the "r potential" for heavy quark binding developing the first (and still the only demonstrable) non-Aeolian gauge theory with an "r" potential; first suggested the existence of topological structures in superfluid He-3; first proposed Yang-Mills theories would appear in condensed matter phenomena with non-scalar order parameters; first developed a grammar-based formalism for quantum computers and applied it to elementary particle theories; first developed a new form of quantum field theory without divergences (thus solving a major 60 year old problem that enabled a unified theory of the Standard Model and Quantum Gravity without divergences to be developed); first developed a formulation of complex General Relativity based on analytic continuation from real space-time; first developed a generalized non-homogeneous Robertson-Walker metric that enabled a quantum theory of the Big Bang to be developed without singularities at t = 0; first generalized Cauchy's theorem and Gauss' theorem to complex, curved multi-dimensional spaces; received Honorable Mention in the Gravity Research Foundation Essay Competition in 1978; first developed a physically acceptable theory of faster-than-light particles; first derived a composition of extremums method in the Calculus of Variations; first quantitatively suggested that inflationary periods in the history of the universe were not needed; first proved Gödel's Theorem implies Nature must be quantum; provided a new alternative to the Higgs Mechanism, and Higgs particles, to generate masses; first showed how to resolve logical paradoxes including Gödel's Undecidability Theorem by developing Operator Logic and Quantum Operator Logic; first developed a quantitative harmonic oscillator-like model of the life cycle, and interactions, of civilizations; first showed how equations describing superorganisms also apply to civilizations. A recent book shows his theory applies successfully to the past 14 years of history and to *new* archaeological data on Andean and Mayan civilizations as well as Early Anatolian and Egyptian civilizations.

He first developed an axiomatic derivation of the form of The Standard Model from geometry – space-time properties – The Unified SuperStandard Model. It unifies all the known forces of Nature. It also has a Dark Matter sector that includes a Dark ElectroWeak sector with Dark doublets and Dark gauge interactions. It uses quantum coordinates to remove infinities that crop up in most interacting quantum field theories and additionally to remove the infinities that appear in the Big Bang and generate inflationary growth of the universe. It shows

gravity has a MOND-like form without sacrificing Newton's Laws. It relates the interactions of the MOND-like sector of gravity with the r-potential of Quark Confinement. The axioms of the theory lead to the question of their origin. We suggest in the preceding edition of this book it can be attributed to an entity with God-like properties. We explore these properties in "God Theory" and show they predict that the Cosmos exists forever although individual universes (or incarnations of our universe) "come and go." Several other important results emerge from God Theory such a functionally triune God. The Unified SuperStandard Theory has many other important parts described in the Current Edition of *The Unified SuperStandard Theory* and expanded in subsequent volumes.

Blaha has had a major impact on a succession of elementary particle theories: his Ph.D. thesis (1970), and papers, showed that quantum field theory calculations to all orders in ladder approximations could not give scaling deep inelastic electron-nucleon scattering. He later showed the eigenvalue equation for the fine structure constant α in Johnson-Baker-Willey QED had a zero at $\alpha = 1$ not 1/137 by solving the Schwinger-Dyson equations to all orders in an approximation that agreed with exact results to 4th order in α thus ending interest in this theory. In 1979 at Prof. Ken Johnson's (MIT) suggestion he calculated the proton-neutron mass difference in the MIT bag model and found the result had the wrong sign reducing interest in the bag model. These results all appear in Physical Review papers. In the 2000's he repeatedly pointed out the shortcomings of SuperString theory and showed that The Standard Model's form could be derived from space-time geometry by an extension of Lorentz transformations to faster than light transformations. This deeper space-time basis greatly increases the possibility that it is part of THE fundamental theory. Recently, Blaha showed that the Weak interactions differed significantly from the Strong, electromagnetic and gravitation interactions in important respects while these interactions had similar features, and suggested that ElectroWeak theory, which is essentially a glued union of the Weak interactions and Electromagnetism, possibly modulo unknown Higgs particle features, be replaced by a unified theory of the other interactions combined with a stand-alone Weak interaction theory. Blaha also showed that, if Charmonium calculations are taken seriously, the Strong interaction coupling constant is only a factor of five larger than the electromagnetic coupling constant, and thus Strong interaction perturbation theory would make sense and yield physically meaningful results.

In graduate school (1965-71) he wrote substantial papers in elementary particles and group theory: The Inelastic E- P Structure Functions in a Gluon Model. Phys. Lett. B40:501-502,1972; Deep-Inelastic E-P Structure Functions In A Ladder Model With Spin 1/2 Nucleons, Phys.Rev. D3:510-523,1971; Continuum Contributions To The Pion Radius, Phys. Rev. 178:2167-2169,1969; Character Analysis of U(N) and SU(N), J. Math. Phys. 10, 2156 (1969); and The Calculation of the Irreducible Characters of the Symmetric Group in Terms of the Compound Characters, (Published as Blaha's Lemma in D. E. Knuth's book: *The Art of Computer Programming Vols. 1 – 4*).

In the early 1980's Blaha was also a pioneer in the development of UNIX for financial, scientific and Internet applications: benchmarked UNIX versions showing that block size was critical for UNIX performance, developing financial modeling software, starting database benchmarking comparison studies, developing Internet-like UNIX networking (1982) and developing a hybrid shell programming technique (1982) that was a precursor to the PERL programming language. He was also the manager of the AT&T ten-year future products development database. His work helped lead to commercial UNIX on computers such as Sun Micros, IBM AIX minis, and Apple computers.

In the 1980's he pioneered the development of PC Desktop Publishing on laser printers and was nominated for three "Awards for Technical Excellence" in 1987 by PC Magazine for PC software products that he designed and developed.

Recently he has developed a theory of Megaverses – actual universes of which our universe is one – with quantum particle-like properties based on the Wheeler-DeWitt equation of Quantum Gravity. He has developed a theory of a baryonic force, which had been conjectured many years ago, and estimated the strength of the force based on discrepancies in measurements of the gravitational constant G. This force, operative in D-dimensional space, can be used to escape from our universe in "uniships" which are the equivalent of the faster-than-light starships proposed in the author's earlier books. Thus travel to other universes, as well as to other stars is possible.

Blaha also considered the complexified Wheeler-DeWitt equation and showed that its limitation to real-valued coordinates and metrics generated a Cosmological Constant in the Einstein equations.

The author has also recently written a series of books on the serious problems of the United States and their solution as well as a book on the decline of Mankind that will follow from current social and genetic trends in Mankind.

In the past twenty years Dr. Blaha has written over 80 books on a wide range of topics. Some recent major works are: *From Asynchronous Logic to The Standard Model to Superflight to the Stars, All the Universe!, SuperCivilizations: Civilizations as Superorganisms, America's Future: an Islamic Surge, ISIS, al Qaeda, World Epidemics, Ukraine, Russia-China Pact, US Leadership Crisis, The Rises and Falls of Man – Destiny – 3000 AD: New Support for a Superorganism MACRO-THEORY of CIVILIZATIONS From CURRENT WORLD TRENDS and NEW Peruvian, Pre-Mayan, Mayan, Anatolian, and Early Egyptian Data, with a Projection to 3000 AD*, and *Mankind in Decline: Genetic Disasters, Human-Animal Hybrids, Overpopulation, Pollution, Global Warming, Food and Water Shortages, Desertification, Poverty, Rising Violence, Genocide, Epidemics, Wars, Leadership Failure.*

He has taught approximately 4,000 students in undergraduate, graduate, and postgraduate corporate education courses primarily in major universities, and large companies and government agencies.

Recently he developed a quantum theory, The Unified SuperStandard Theory (UST), which describes elementary particles in detail without the difficulties of conventional quantum field theory. He found that the internal symmetries of this theory could be exactly derived from a 32 dimension complex quaternion theory called QUeST. He further found that a 32 dimension complex octonion theory (MOST) describes the Megaverse. It can hold QUeST universes such as our own universe. It has an internal symmetry structure which is a superset of the QUeST internal symmetries.

www.ingramcontent.com/pod-product-compliance
Lightning Source LLC
Chambersburg PA
CBHW082008190326
41458CB00010B/3115